AF738354

Carbon Dots

Ya-Ping Sun

Carbon Dots

Exploring Carbon at Zero-Dimension

Springer

Ya-Ping Sun
Department of Chemistry
Clemson University
Clemson, SC, USA

ISBN 978-3-030-41183-1 ISBN 978-3-030-41184-8 (eBook)
https://doi.org/10.1007/978-3-030-41184-8

© Springer Nature Switzerland AG 2020
This work is subject to copyright. All rights are reserved by the Publisher, whether the whole or part of the material is concerned, specifically the rights of translation, reprinting, reuse of illustrations, recitation, broadcasting, reproduction on microfilms or in any other physical way, and transmission or information storage and retrieval, electronic adaptation, computer software, or by similar or dissimilar methodology now known or hereafter developed.
The use of general descriptive names, registered names, trademarks, service marks, etc. in this publication does not imply, even in the absence of a specific statement, that such names are exempt from the relevant protective laws and regulations and therefore free for general use.
The publisher, the authors, and the editors are safe to assume that the advice and information in this book are believed to be true and accurate at the date of publication. Neither the publisher nor the authors or the editors give a warranty, expressed or implied, with respect to the material contained herein or for any errors or omissions that may have been made. The publisher remains neutral with regard to jurisdictional claims in published maps and institutional affiliations.

This Springer imprint is published by the registered company Springer Nature Switzerland AG
The registered company address is: Gewerbestrasse 11, 6330 Cham, Switzerland

Acknowledgment

The author thanks all former and current students, postdoctoral fellows, research scientists, visiting scholars, and other associates in the Sun research group at Clemson University and also internal and external research collaborators for their participation in and contributions to the investigation of carbon dots and related nanomaterials.

The author and the Sun research group at Clemson University gratefully acknowledge the generous financial support of the research on carbon dots and related nanomaterials by the National Science Foundation (NSF), National Institutes of Health (NIH), Air Force Office of Scientific Research (AFOSR), Air Force Research Laboratory (AFRL), US Department of Agriculture (USDA), DoD Breast Cancer Research Program (USAMRMC), *Susan G. Komen for the Cure* Breast Cancer Foundation, Environmental Protection Agency (EPA), and National Aeronautics and Space Administration (NASA) and the South Carolina Space Grant Consortium.

The preparation of this book was made possible by the support of AFOSR (through the program of Dr. Kenneth Caster) and NSF.

Contents

Chapter 1
Introduction

Carbon dots (CDots, Fig. 1.1) (Sun 2010; Sun et al. 2006; Cao et al. 2007) (also referred to as "carbon quantum dots," "carbon nanodots," "carbon nano-dots," "C-Dots," "C-dots," "CDs", etc.) are generally defined as small carbon nanoparticles with various surface passivation schemes (Cao et al. 2013). Since the first report in 2006 (Sun et al. 2006), CDots have attracted a tremendous amount of attention around the world, as reflected by the large and ever-increasing number of publications in the literature. According to a simple search of citation database, there have already been around 3000 publications directly or indirectly related to CDots (Yang and Ho 2019). In fact, the research and development of CDots have emerged to represent a rapidly advancing and expanding technical field.

The discovery and subsequent development of CDots, especially in earlier years, were driven by the need for high performance in terms of optical absorption and fluorescence emission properties (Fig. 1.2) yet nontoxic alternatives to conventional semiconductor quantum dots (QDs), which are mostly semiconductor nanocrystals of physical dimensions smaller than the exciton Bohr radius for the quantum confinement effect. In this regard, CDots have already done exceedingly well, with their known performance in terms of optical properties competitive to that of the more established semiconductor QDs, such as the commercially available CdSe/ZnS core-shell nanoparticles, yet benign and nontoxic (Yang et al. 2009). Beyond that, CDots offer unique and/or advantageous properties valuable to a number of technological applications. Among some representative examples are bioimaging in vitro and in vivo, antibacterial and antiviral functions, bio-delivery, sensing, optoelectronic materials and devices, and others. Interestingly and somewhat surprisingly, many of the known characteristics of CDots resemble not only phenomenologically but also to a significant extent mechanistically those of conventional semiconductor QDs. Such property and mechanistic similarities or "convergence" for the seemingly completely different materials but both at the nanoscale may have far-reaching implications beyond the understanding and further development of CDots and related technologies. The dramatic changes in electronic properties from bulk carbon to nanoparticles, similar to those famously found between gold and gold

© Springer Nature Switzerland AG 2020
Y.-P. Sun, *Carbon Dots*, https://doi.org/10.1007/978-3-030-41184-8_1

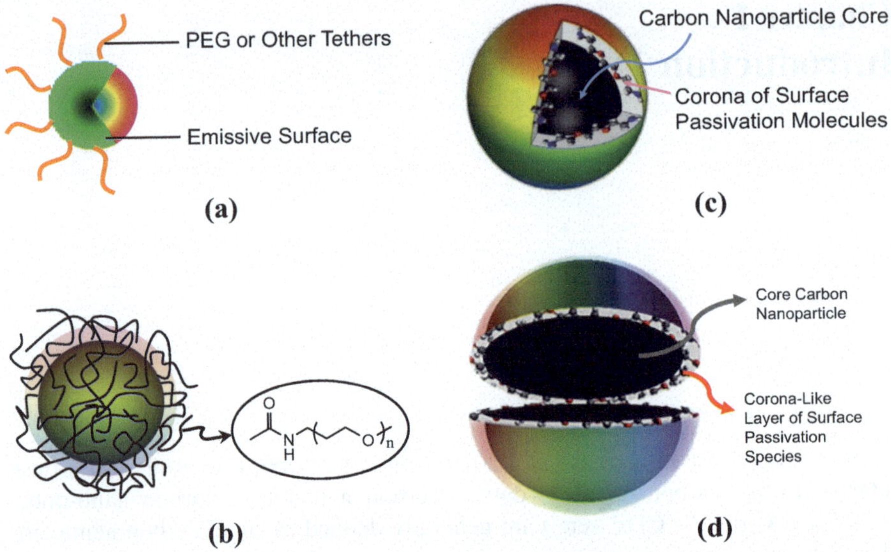

Fig. 1.1 Various cartoons used by the author and associates to illustrate the structure of CDots: **a** in the original report (Sun et al. 2006), **b** a few years later (e.g., Wang et al. 2010a), and **c** and **d** in more recent publications. All of the illustrations emphasize photoluminescence emissions from the passivated surface of carbon nanoparticles, a proposed rationalization of CDots in the original report that has generally been consistent with the available experimental results

nanoparticles, demonstrate yet again the unique characteristics of materials at the nanoscale. One may extrapolate to wonder if the same dramatic bulk-to-nano property transformations could be found in similar materials of other elements.

CDots realize and enhance the properties intrinsic to small carbon nanoparticles with the particle surface passivation in various schemes (Fig. 1.1). The carbon nanoparticles are distinctively different in structures and also in a number of properties from other widely pursued nanoscale carbon allotropes, from zero-dimensional fullerenes to one-dimensional carbon nanotubes and more recently two-dimensional graphenes (Fig. 1.3), though the nanoparticles have in many ways been underappreciated or even largely ignored by the research community. However, the small carbon nanoparticles should in fact be considered as representing a distinct zero-dimensional carbon allotrope, at least sharing the title with fullerenes as one may further argue that fullerenes are stoichiometrically and electronically singularly defined large molecules, hardly nanoscale materials. The unique and/or advantageous properties of small carbon nanoparticles become evident as a result of the substantial enhancements in the configuration of CDots, as to be highlighted throughout this book.

The small carbon nanoparticles are populated by surface defects, which may serve as traps for electrons and holes following photoinduced charge transfers in the nanoparticles. The surface passivation in various schemes likely serves the function of stabilizing the defect-derived traps to facilitate more efficient radiative

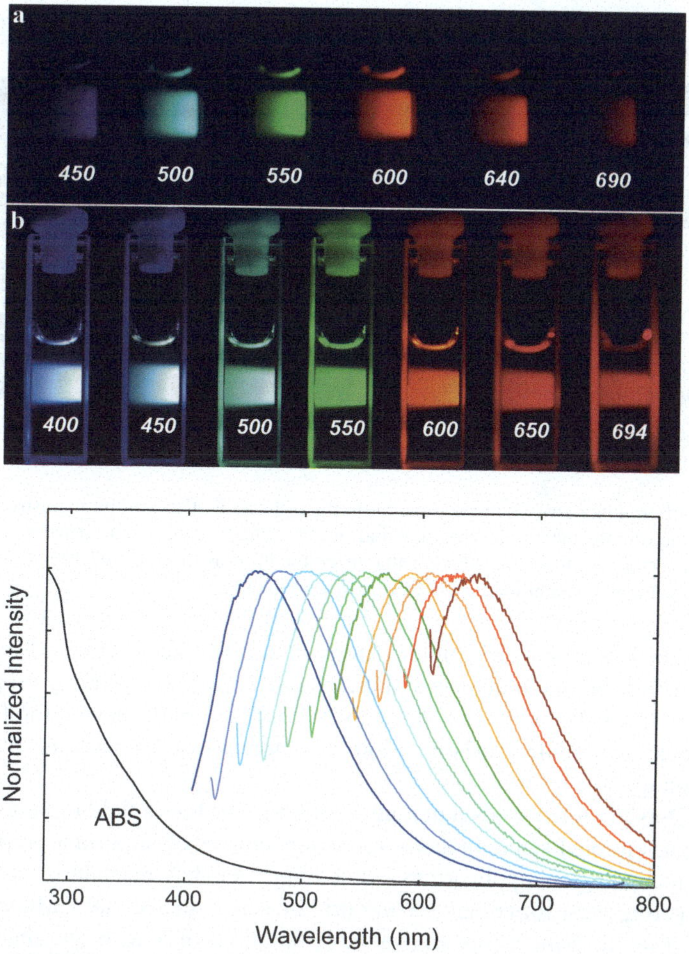

Fig. 1.2 (Upper) Aqueous solutions of the oligomeric polyethylene glycol diamine (PEG_{1500N})-functionalized CDots (a) excited at 400 nm and photographed through band-pass filters of different wavelengths as indicated and (b) excited at the indicated wavelengths and photographed directly. (Lower) The absorption spectrum (ABS) and photoluminescence emission spectra (with progressively longer excitation wavelengths from 400 nm on the left in 20 nm increment) of poly(propionylethylenimine-*co*-ethylenimine) (PPEI-EI)-functionalized CDots in an aqueous solution. (Originally Figures 1 and 3 in Sun et al. (2006). Reproduced with permission)

recombinations of the trapped electrons and holes, as reflected by the observed bright photoluminescence emissions, and also other associated photoexcited state processes, such as those responsible for redox activities and photodynamic effects. The same kinds of defects are apparently also popular in other nanoscale carbon allotropes including carbon nanotubes and graphenes (Fig. 1.3). The passivation of those defects in the same or similar schemes has led to the observation of largely the same bright and colorful photoluminescence emissions from the corresponding car-

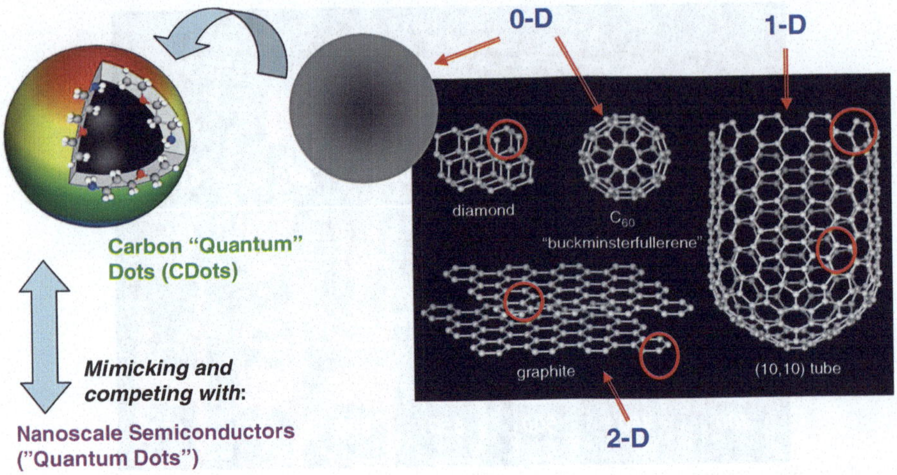

Fig. 1.3 Carbon nanoparticles are compared on the right with other carbon nanostructures (with the circles marking the inevitable structural and surface defects in those nanostructures) and on the left for their serving as the cores in CDots that resemble in some mechanistic aspects and compete in performance with conventional nanoscale semiconductors

bon nanotube and graphene samples (Riggs et al. 2000; Veca et al. 2009). Thus, the notion that "it is all about defects and their passivation," as originally proposed and justified by Sun and coworkers (Cao et al. 2013), deserves further exploration across the various nanoscale carbon allotropes toward a consistent mechanistic understanding.

The CDots platform holds a unique position in the broad field of nanomaterials, rivaling conventional semiconductor QDs in a number of important properties and potential technological applications yet sharing some key mechanistic characteristics, competing with fullerenes for the title of zero-dimensional carbon allotrope, and exhibiting the critical role and valuable functions of defects and their effective passivation, which also represent the similar behaviors found in the other nanoscale carbon allotropes. These attributes have stimulated considerable interest in the development of CDots-derided or related technologies. In fact, the application targets are extremely broad, across almost all technological areas covered by nanoscale semiconductors and other carbon nanomaterials. With all these extremely broad and active efforts, commercial development and uses of CDots in not so distant future may be envisaged.

In this book, small carbon nanoparticles for their structural characteristics and distinctive optical and other properties are highlighted in reference to other carbon nanomaterials, in support of the assertion by the author that the nanoparticles, more than fullerene molecules, should be granted the title of zero-dimensional carbon allotrope. How their intrinsic properties are exploited and enhanced in CDots via the effective passivation of the abundant defects in the nanoparticles is discussed, along with a historical account on the discovery of CDots. Described in more detail are the various synthetic approaches that have been employed for the preparation of CDots and their pros and cons (including especially some serious issues with the poorly synthesized dot samples in many reported studies), the generally consistent structural and property characteristics of CDots, and some representative potential uses of the dots in various configurations and their derived hybrid nanostructures. Some perspectives on the excellent opportunities and also challenges in the further development and mechanistic understanding of the CDots platform are provided. An important note is that the author purposely tries not to have this book become another "comprehensive" review, but rather to present the available information in the related research fields primarily or exclusively from the author's perspective. Readers who are more interested in different presentations and/or views are encouraged to pick from the list of review articles provided in the bibliography.

Chapter 2
Nanoscale Carbon Allotropes

Carbon at the nanoscale has been a major cornerstone in the extremely broad and active research field of nanoscience and nanotechnology. In fact, the dawn in this field coincided with the discovery of fullerene C_{60} in 1985 and the award of Nobel Prize in Chemistry in 1986 for the discovery. Fullerenes have since been widely recognized as "carbon nanomaterials," despite the fact that they are stoichiometrically defined molecules of distinct electronic structures, thus different from other famous nanomaterials such as nanoscale semiconductors. Nevertheless, the discovery did mark the creation of a new carbon allotrope, commonly regarded as an allotrope at the zero-dimension that is very different from the traditional carbon allotropes such as diamond and graphite. The subsequent discovery of carbon nanotubes, both single-walled and multiple-walled (including double-walled), created a new class of carbon nanomaterials that are structurally one-dimensional. These nanotubes are actual nanomaterials, not large tube-shaped all-carbon molecules, as made evident by their varying structural details such as lengths and diameters and especially the inevitable presence of defects. Graphenes, the two-dimensional carbon allotrope, represent another new class of carbon nanomaterials. Again, these are nanoscale materials, not large π-conjugated molecules to be extrapolated all the way down to benzene as what have been implied in some publications. In fact, a clear distinction between "large molecules" and "nanomaterials" is critical to the study and understanding of the various nanoscale carbon allotropes, which will be recapped in the discussion in various sections throughout this book.

On small carbon nanoparticles serving as the cores in CDots (Fig. 1.1), the "small" is relative. These nanoparticles are generally somewhat larger than fullerenes (which are 1–2 nm in diameter) but much smaller than particles in carbon black (which are typically >100 nm to microns in diameter). In fact, the overwhelming majority of the reported CDots have the core carbon nanoparticles of less than 10 nm in diameter. The consequence of larger or substantially larger carbon nanoparticles for CDots-like configurations is mostly unknown. Thus, the small carbon nanoparticles in the general definition on CDots (Fig. 1.1) are those of sizes similar to or somewhat larger than fullerenes. These are the carbon nanoparticles that

© Springer Nature Switzerland AG 2020

Y.-P. Sun, *Carbon Dots*, https://doi.org/10.1007/978-3-030-41184-8_2

represent another (or actually the true, as one might argue) zero-dimensional carbon allotrope and also a distinct class of carbon nanomaterials, in contrast to zero-dimensional fullerenes as stoichiometrically and electronic structure-wise well-defined molecules.

1 Fullerenes Versus Carbon Nanoparticles

Fullerenes and small carbon nanoparticles (Fig. 2.1) may both be considered as zero-dimensional carbon allotropes, though the former are actually "large molecules" for their defined molecular and electronic structures, and the latter are "small materials" for their variations in structures or even "compositions" and more importantly the presence of structural and/or surface defects. Here compositions refer to the relative contents of amorphous carbon and graphitic crystals in the nanoparticles. Available experimental results have not provided any convincing evidence for any significant effects of the compositions in carbon nanoparticles on their properties that are relevant to CDots. In fact, most results suggest generally insensitivity of the observed optical and related properties of carbon nanoparticles to their sources and processing history.

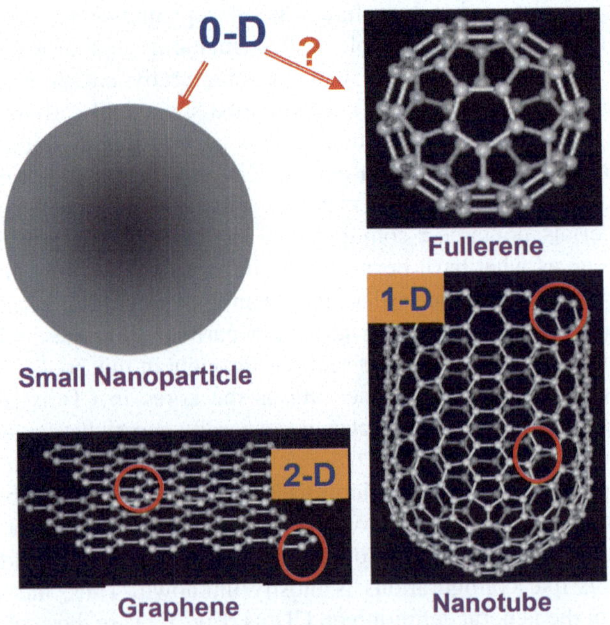

Fig. 2.1 Carbon nanomaterials in different dimensions, with the red circles marking some areas for potential structural and edge defects

1.1 Fullerenes and Derivatives

As highly symmetric all-carbon molecules, fullerenes have many unique properties, which have been widely explored for fundamental understanding and technological applications (Hirsch and Brettreich 2005; Fowler and Manolopoulos 2006). Among those more relevant to carbon nanoparticles and CDots are properties related to uses in optoelectronic devices and systems, namely, optical and charge-transfer properties. In this regard, one might find it somewhat surprising that fullerenes would be chosen for anything optical at all, because parent fullerenes and their functionalized derivatives are neither strong photon absorbers nor bright emitters. For C_{60}, the optical absorptions in the visible are rather weak, with the peak molar absorptivity at around 530 nm (Fig. 2.2) of only ~950 M^{-1} cm^{-1} (Ma and Sun 1996) or normalized to the molar concentration of carbon (M_{Carbon}) by dividing 60 to 16 $M_{Carbon}{}^{-1}$ cm^{-1} (significantly lower than that of small carbon nanoparticles, as compared later in the chapter). C_{70} is only slightly more absorptive in the same spectral region due to its reduced molecular symmetry (Fig. 2.2).

Fullerenes are also weak emitters. Fluorescence emissions from C_{60} in a solution can be detected but only on a more sensitive spectrometer (Fig. 2.2). Since the overwhelming majority of commercial emission spectrometers are much less sensitive in the longer wavelength spectral region (progressively weaker from red to near-IR), the observed fluorescence emission spectra of fullerenes must be corrected very carefully for such nonlinear instrumental responses in the individual spectrometers. With the use of a more red/near-IR sensitive emission spectrometer coupled with the corrections for nonlinear instrumental responses, the fluorescence quantum yield of C_{60} in a toluene solution was determined as 0.032% (Ma and Sun 1996), which has since been verified and/or adopted by many other research groups around the world. The observed fluorescence quantum yields of C_{60} in various other solvents are generally around the value determined in toluene (Table 2.1), exhibiting no major solvent dependence.

Fullerenes can readily be functionalized in a number of chemical reactions, such as those that have been particularly popular for the preparation of methano-fullerene and pyrrolidino-fullerene derivatives. However, the corresponding C_{60} derivatives are in general only slightly more absorptive than the parent C_{60}, as concluded on the basis of quantitative optical spectroscopic investigations on a series of derivatized C_{60} molecules (Fig. 2.3) (Sun et al. 2000). For example, the optical absorption features shared by methano-C_{60} derivatives include a characteristic weak band around 700 nm (Fig. 2.4), with molar absorptivities on the order of 200–300 M^{-1} cm^{-1} (Ma et al. 1997; Sun et al. 1998, 2000). The famous PCBM, or 1-(3-(methoxycarbonyl) propyl)-1-phenyl-[6,6]C61, which has been the C_{60} derivative of choice in optoelectronic devices, is a methano-C_{60} molecule (the compound **5** in Fig. 2.3). Its optical absorptivities are still on the same order of magnitude as those of the parent C_{60}, despite the spectral changes associated with the derivatization (Fig. 2.4, Table 2.2) (Sun et al. 1998, 2000). Again, fullerenes are in general not so efficient in harvesting the visible portion of the solar irradiation, representing a major limitation in some of the energy conversion applications.

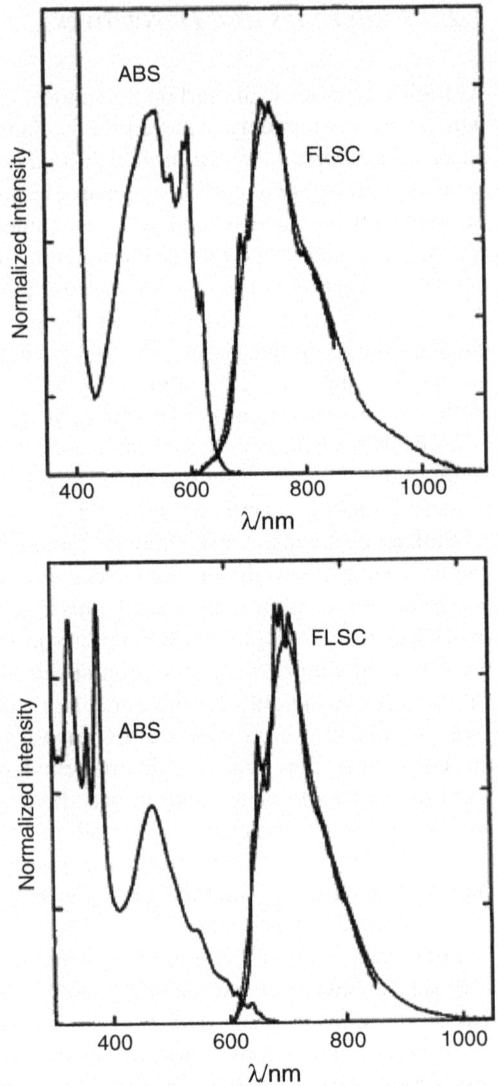

Fig. 2.2 Absorption (ABS) and fluorescence (FLSC) spectra of C_{60} (upper) and C_{70} (lower) in room-temperature hexane. The fluorescence spectra were obtained by using a regular detector and small slit (——) and a near-infrared-sensitive detector and large slit (-·-·-·-). (Originally Figure 2 and Figure 3 in Ma and Sun (1996). Reproduced with permission)

Fullerene derivatives from organic functionalization reactions are similarly weak in fluorescence emissions, namely, that the lack of any major effects on optical absorptivities due to the reduced molecular symmetry also applies to the corresponding fluorescence emissions. For C_{60} mono-adducts (e.g., Fig. 2.3), the emission bands are shifted to longer wavelengths beyond 700 nm (Fig. 2.5) (Ma et al. 1997, Sun et al. 1998, 2000).

Adducts represent most of the fullerene derivatives, such as the mono-adducts in a series of methano-C_{60} and pyrrolidino-C_{60} derivatives highlighted above. Since fullerene cages are capable of accepting multiple additions, similar organic functionalization reactions could produce bis- and multi-adducts (e.g., Fig. 2.3).

Table 2.1 Observed fluorescence quantum yields of C_{60} and C_{70} and yield ratios ($\Phi_{F,C70}/\Phi_{F,C60}$) under different solvent conditions

Solvent	*n*	*ε*	$\Phi_{F,C60}(10^4)$	$\Phi_{F,C70}(10^4)$	$\Phi_{F,C70}/\Phi_{F,C60}$
Hexane	1.375	1.890	3.3	5.9	1.8
Dichloromethane	1.424	9.080	3.3	5.4	1.6
Chlorobenzene	1.524	5.710	2.8	5.3	1.9
o-Dichlorobenzene	1.552	9.930	3.0	4.8	1.6
Carbon disulfide	1.632	2.641	2.6	4.2	1.6
Toluene	1.496	2.379	3.2	5.7	1.8
o-Xylene	1.506	2.568	2.9	5.7	2.0
1,2,4-Trimethylbenzene	1.505		3.1	5.8	1.9
1,2,3,5-Tetramethylbenzene	1.513		3.2	5.7	1.8
1-Methylnaphthalene	1.618	2.920	3.7	6.2	1.7

Originally Table 1 in Ma and Sun (1996). Reproduced with permission

However, regardless of the number of additions and the additions being symmetric or nonsymmetric on the fullerene cage in multi-adducts, these adducts are generally still weak optical absorbers and emitters. In fact, for the methano-C_{60} tris-adduct **10** in Fig. 2.3 as an example, the fluorescence emissions are even weaker than those of the parent C_{60} and corresponding mono-adduct (Table 2.3).

The further push in the functionalization reactions could produce fullerene hexakis-adducts. These molecules are structurally rather pretty, but their optical spectroscopic properties generally offer no major advantages over those of other adducts, even for those molecules in which the fullerene cage is coupled with 12 pyrene species (Fig. 2.6) (Sun et al. 2000; Martin et al. 2003; Li et al. 2005b). Somewhat interesting for these symmetrically pyrene-tethered hexakis-adducts were the observed simple mono-exponential decays of fluorescence emissions with excitations into the pyrene absorption bands, including the pyrene excimer emissions (Fig. 2.7). The fullerene cage at the center in these molecules behaved as a quencher for the pyrene fluorescence emissions, which might be expected due to excited state energy transfers (Martin et al. 2003). No other major effects on the absorption and fluorescence emission characteristics of the hexakis-added fullerene cage in these pyrene-loaded molecules beyond those typical for other hexakis-adducts (e.g., **11** and **12** in Fig. 2.3) have been observed, including results obtained in follow-up investigations of these pretty molecules.

The interest in fullerenes for uses in optoelectronics may be attributed to their excellent electron acceptor characteristics in the ground state and more so in the photoexcited states. Fullerenes are known to form charge-transfer complexes with strong electron donors such as *N*,*N*-dialkylanilines, which exhibit the characteristic broad optical absorption features at the longer wavelength side of the fullerene absorptions (e.g., Fig. 2.8 (Wang 1992)). Fluorescence emissions of fullerenes are readily quenched by electron donor molecules, including the formation of emissive exciplexes with some of the donor molecules (Fig. 2.9 (Sun et al. 1994)), which are complexes in the photoexcited states only. Since fullerene cages are readily functionalized in a number of established reaction schemes, there have been extensive investigations on the design, synthesis, and spectroscopic and mechanistic

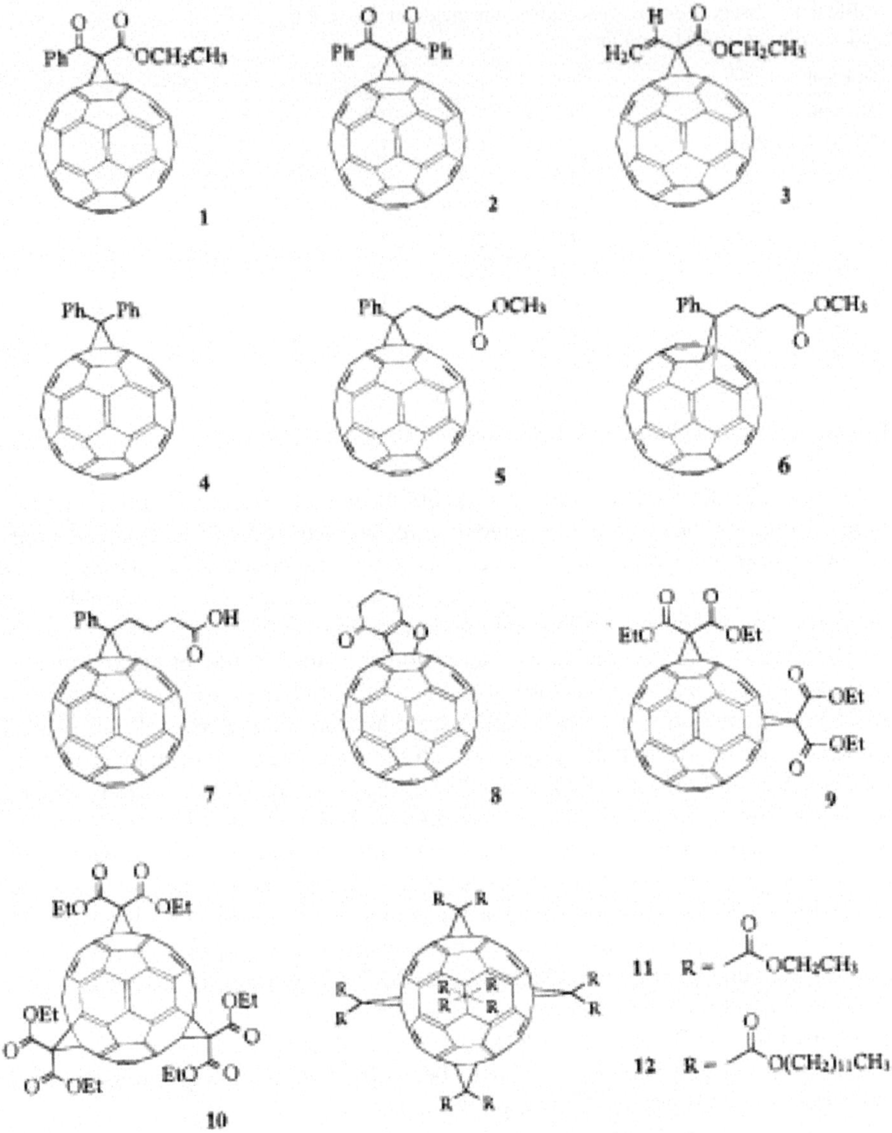

Fig. 2.3 Selected C_{60} mono-adducts and multi-adducts. (Sun et al. 2000)

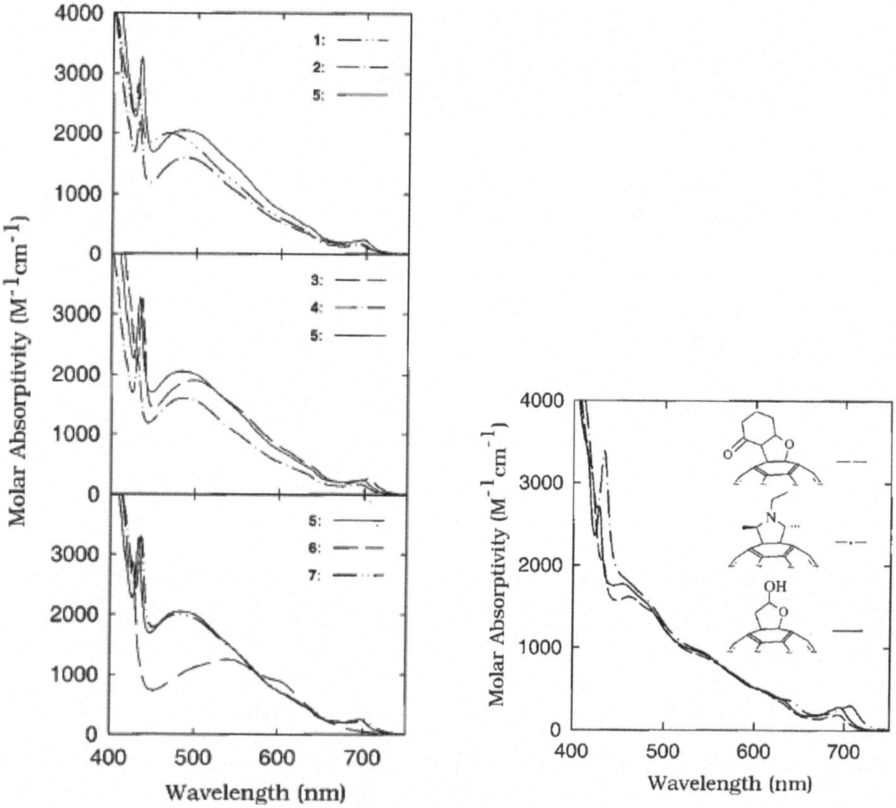

Fig. 2.4 Absorption spectra of **1**–**3** and **5**–**7** in toluene and **4** in CS_2 at room temperature (left) and **8** in room-temperature toluene compared with those of the C_{60} derivatives functionalized with a 5-member ring (right). The compound numbers correspond to those in Fig. 2.3. (Originally Figure 1 and Figure 2 in Sun et al. (2000). Reproduced with permission)

characterizations of fullerene-derived redox dyads, triads, and related molecular architectures (e.g., Fig. 2.10). The study of intramolecular electron transfers in these systems has been an extremely popular topic in the literature.

It was established very early in the exploration of fullerenes that a C_{60} or C_{70} cage could accept up to six electrons (Xie et al. 1992), which served to highlight the strong electron acceptor characteristics of the fullerene molecules. Since then, there have been efforts on the design, synthesis, and exploration of fullerene derivatives containing multiple electron donors. For example, the strongly electron-donating *N*,*N*-dimethylaniline (DMA) units were attached to a C_{60} cage in different numbers, up to six in a symmetric tris-adduct (Fig. 2.11 (Sun et al. 2002b; Li et al. 2005a)). The purpose was to probe the intramolecular transfer of potentially multiple electron equivalents from the donors to the fullerene cage. The results did show ground-state charge transfers in the various "**D**" molecules, with the structurally comparable "**R**" molecules as references (Fig. 2.11). The charge-transfer absorption features could be

Table 2.2 Absorption spectral parameters of the compounds in Fig. 2.3

Compound	Solvent	abs_{0-0} (nm)	ϵ_{0-0} (M^{-1} cm^{-1})	abs_{max} (nm)	ϵ_{max} (M^{-1} cm^{-1})
1	Toluene	690	256	467	2000
2	Toluene	694	173	487	1600
3	CS_2	694	260	487	1970
4	Toluene	705	274	497	1900
5	Toluene	697	246	483	2050
6	Toluene			538	1250
7	CS_2	703	285	480	2390
8	Toluene	687	136	458	1070
9	Toluene			482	2790
10	Toluene			477	2930
11	Toluene	[a]		~40 at 500 nm	
12	Hexane			539(sh)	443

[a]No peak in the visible region
Originally Table 1 in Sun et al. (2000). Reproduced with permission

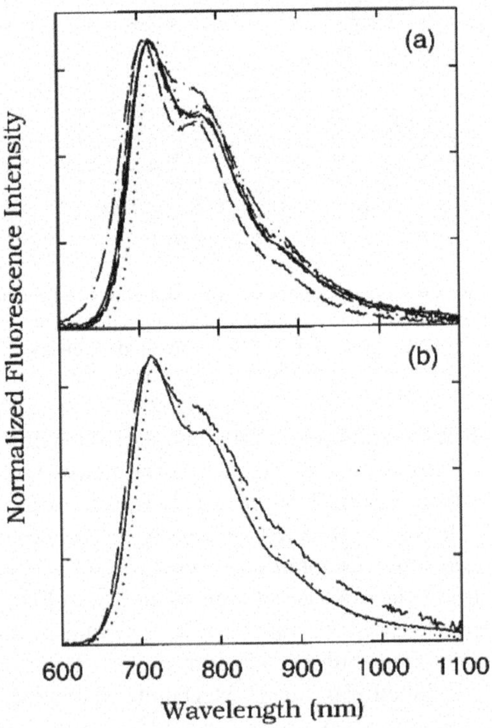

Fig. 2.5 Fluorescence spectra of (a) **1** (-·-·-·-), **2** (-·-·-·-), **4** (······), **5** (——), and **8** (- - - -) and (b) **5** (——), **6** (- - - -), and **7** (······) in toluene at room temperature. The compound numbers correspond to those in Fig. 2.3. (Originally Figure 4 in Sun et al. (2000). Reproduced with permission)

Table 2.3 Fluorescence properties of the compounds in Fig. 2.3

Compound	Solvent	$FLSC_{0-0}$ (nm)	$\Phi_F \times 10^{-3}$	τ_F (ns)
1	Toluene	707	1.3	1.4
2	Toluene	714	0.85	1.5
3	Toluene	716	1.2	1.4
4	CS_2	715	1.2	1.3
5	Toluene	715	1.4	1.4
6	Toluene	712	0.4	1.2
7	CS_2	722	1.5	1.3_5
8	Toluene	706	1.4	1.8
9	Toluene	780[a]	1.5	2.9_5
	CH_2Cl_2			
10	Toluene	723[a]	0.8	2.4_5
	CH_2Cl_2		0.7	1.9
11	Toluene	694[a]	2.4	2.1
	CH_2Cl_2		2.3	1.8_5
	Methanol			2.9_5
12	Hexane	664[a]	1.8	2.0

[a]Band maximum

Originally Table 2 in Sun et al. (2000). Reproduced with permission

diminished by the protonation of the DMA units with trifluoroacetic acid (TFAA), as illustrated in Fig. 2.12 (Li et al. 2005a). However, while the contributions of intramolecular charge-transferred species to the observed absorptivities were found to be different in the series of **D** compounds, generally increasing with the number of DMA units (Table 2.4), a more quantitative examination of the intramolecular charge-transfer equilibrium suggested the equilibrium constant only varying slightly from $\mathbf{D_2}$ to $\mathbf{D_6}$. Thus, "an inevitable conclusion from these results and analyses is that the ground-state intramolecular charge transfer 'saturates' with only two electron-donating DMA units in the methano-C_{60} adducts" (Li et al. 2005a). Further, "the fluorescence results suggest that the photoinduced electron transfers in the **D** series are strongly dependent on the medium polarity. The **D** molecules in a polar solvent like methylene chloride are essentially dark (nonemissive). Because the aromatic amines are the electron donors in these molecules, their protonation in the presence of an organic acid in the solutions may result in a complete elimination of the intra-molecular electron-transfer processes in both ground and excited states." The acid effect was found to be reversible without any damages to the **D** compounds.

A methano-C_{60} hexakis-adduct symmetrically tethered with 12 electron-donating DMA units was synthesized (Fig. 2.13) (Li et al. 2006). Despite the higher number of electron donors and the presence of "spacers" between DMA units and the fullerene cage (making the configuration of individual DMA-cage pair similar to a dyad), experimentally there was no evidence for the targeted "injection of multiple electron equivalents" into the fullerene cage. However, an interesting twist was that apparently the molecule could readily be "visualized" in atomic force microscopy (AFM) imaging (Fig. 2.13).

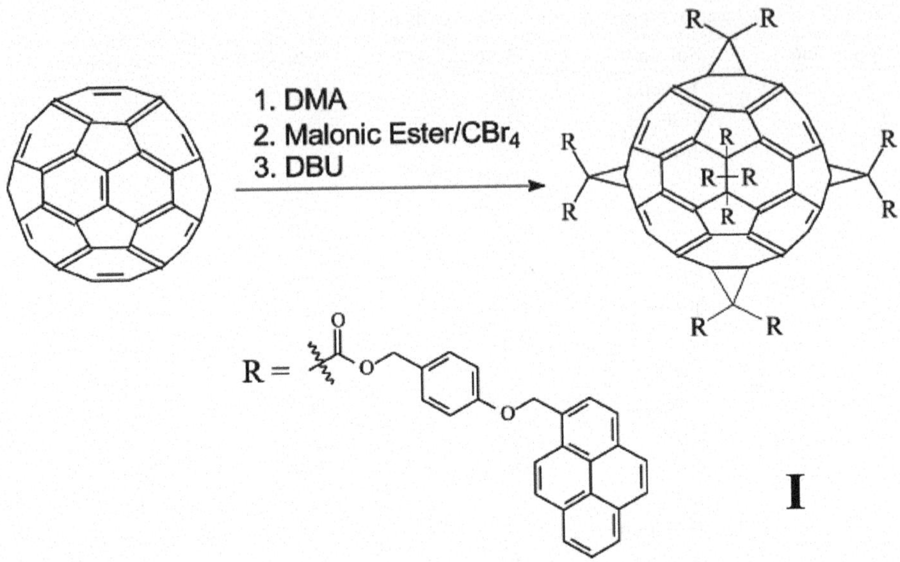

I

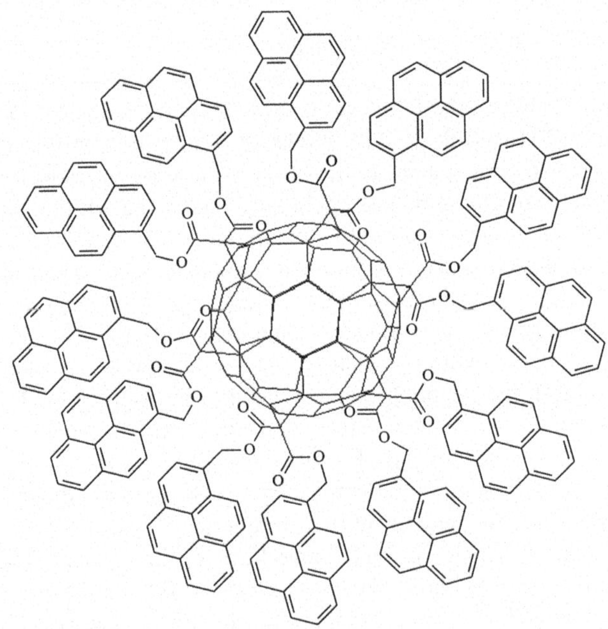

II

Fig. 2.6 Methano-C_{60} hexakis-adducts with symmetrically tethered 12 pyrene units. (Martin et al. 2003; Li et al. 2005b)

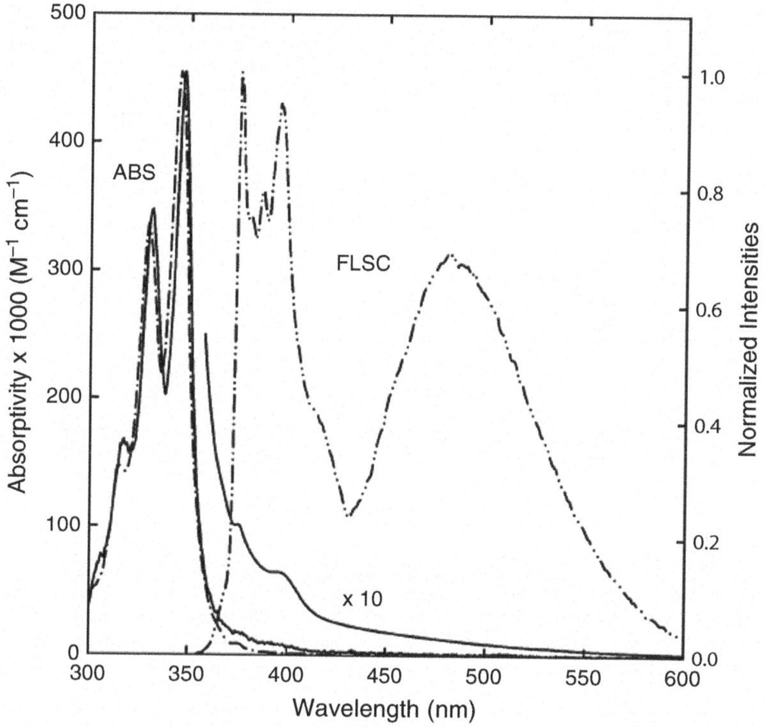

Fig. 2.7 Absorption (ABS, ——), fluorescence (FLSC, 330 nm excitation, -·-·-·-), and fluorescence excitation (490 nm emission, -·-·-·-) spectra of the hexakis-adduct **I** shown in Fig. 2.6 in room-temperature chloroform solution. (Originally Figure 1 in Martin et al. (2003). Reproduced with permission)

Fullerenes and their derived materials have been so extensively investigated, for which there have already been many authored and edited books. Therefore, the descriptions above are obviously not intended for a comprehensive review, far from it, but rather for the purpose of setting the stage for comparisons between fullerenes and small carbon nanoparticles. Beyond the difference between the molecular nature of the former and the nanoscale materials for the latter, the highlighted optical and redox properties of fullerenes are more relevant to the focus of this book, as most of the research and development of CDots have been on the exploration and exploitation of their effective photon harvesting, competitive photoluminescence performance, and unique and/or advantageous photoinduced redox characteristics.

1.2 Small Carbon Nanoparticles

"Small" here refers to the general requirement for the carbon nanoparticles to be small enough that would structurally and property-wise qualify their representing a distinct zero-dimensional carbon allotrope. Practically as related to their serving as

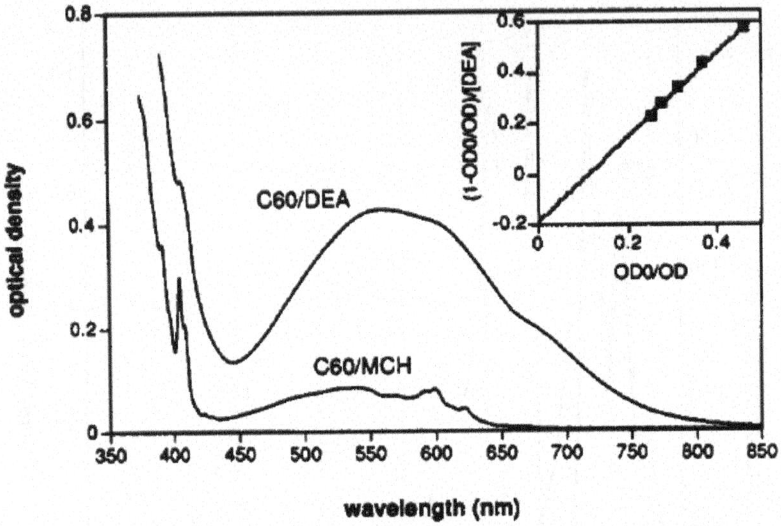

Fig. 2.8 Absorption spectra of C_{60} in methylcyclohexane (C_{60}/MCH) and *N*,*N*-diethylaniline (C60/DEA). The insert shows a plot of (1 – OD_0/OD)/[DEA] vs OD_0/OD, where OD is the optical density of C_{60} absorption at 600 nm in the presence of DEA and OD_0 in the absence of DEA. The intercept of the best-fit straight line gives the negative equilibrium constant for the C_{60}/DEA charge-transfer complex. (Originally Figure 1 in Wang (1992). Reproduced with permission)

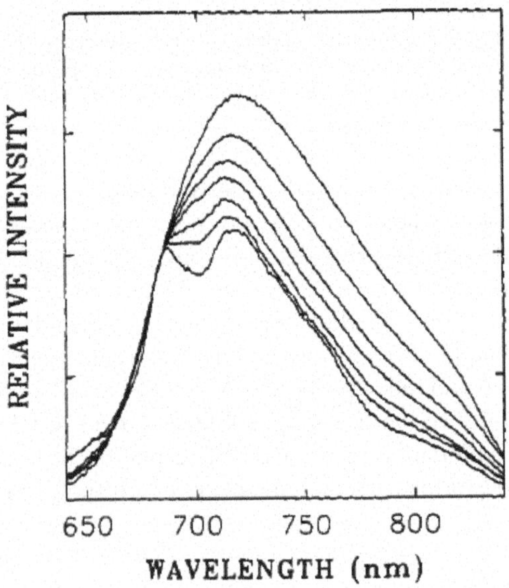

Fig. 2.9 Observed fluorescence spectra of C_{60} (5×10^{-5} M) in room-temperature hexane at different *N*,*N*-diethylaniline (DEA) concentrations. The DEA concentrations for the spectra (in the order of increasing intensity at 720 nm) are 0, 0.013, 0.025, 0.050, 0.075, 0.126, and 0.176 M. (Originally Figure 7 in Sun et al. (1994). Reproduced with permission)

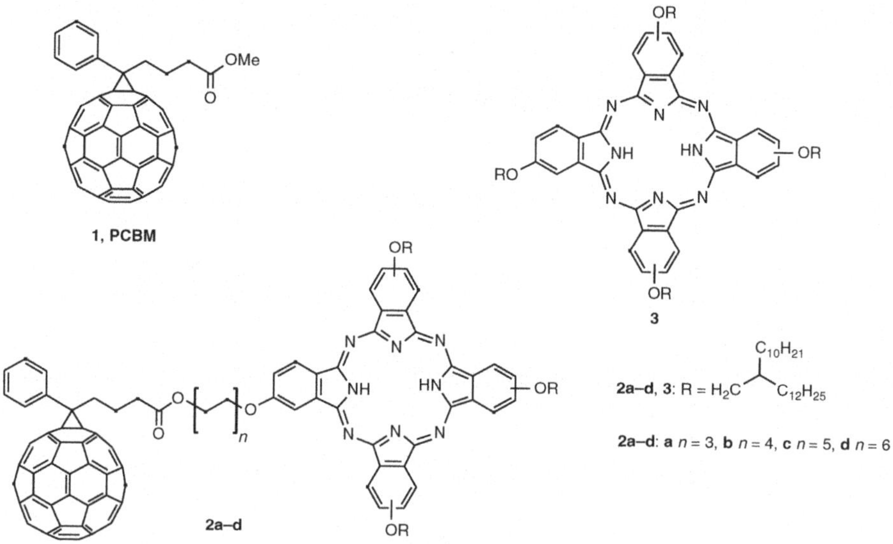

Fig. 2.10 Phthalocyanine-C_{60} dyads related to PCBM, as reported in Geerts et al. (2009)

the carbon cores in CDots, mostly carbon nanoparticles of sizes on the order of 10 nm or less in diameter have been employed. The nanoparticles of smaller sizes are more dispersible in aqueous media and organic solvents, and they are generally associated with a higher population of surface defects, both of which are favorable to the preparation of high-quality CDots. However, there may be a technical limit on how small the nanoparticles can be while they still remain stable, especially for surviving the harsh oxidative conditions that are often used in the processing of or for the carbon nanoparticles (Sun et al. 2006; Sun 2010). Empirical experimental evidence seems to suggest that the lower end in the nanoparticle size distributions is likely somewhere around 2 nm in diameter for small carbon nanoparticles processed and harvested in protocols involving oxidative acid treatments.

There are a number of carbon precursors from which small carbon nanoparticles can be sourced. These are mostly carbon soot samples produced by various established techniques, including laser ablation of carbon targets, the arc discharge of two carbon rods in a chamber similar to the reactor used for fullerenes but with different operational parameters, the carbonization of carbon-rich organic materials at high temperatures under inert conditions, and the breakdown of selected carbon materials into the nanoscale.

1.2.1 Laser Ablation

In a typical setup (Fig. 2.14) (Suda et al. 2002; Sun et al. 2006), a carbon target (which can be prepared, for example, by hot pressing a mixture of graphite powder and carbon-based cement, followed by stepwise baking, curing, and annealing in

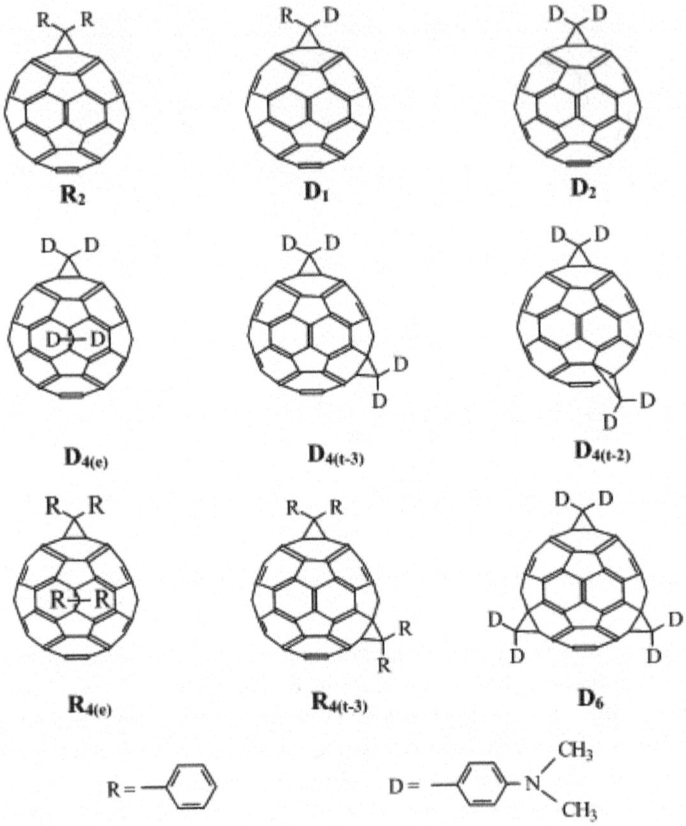

Fig. 2.11 Methano-C_{60} adducts with up to six electron donor (D) units. (From Li et al. (2005a). Reproduced with permission)

argon flow) is placed in a quartz tube fitted in a tube furnace at high temperatures for the ablation by a laser of sufficient energy, such as a Q-switched Nd:YAG laser (1064 nm, 10 Hz). There was some evidence that the ablation in the presence of water vapor could facilitate the production of smaller carbon nanoparticles. Experimentally this could be achieved by purging the ablation chamber with a flow of argon gas carrying water vapor (through a water bubbler). The sample thus produced generally has the appearance of a typical carbon soot sample. Scanning electron microscopy images exhibit features consistent with the presence of abundant nanoparticles (Fig. 2.15).

The laser ablation method and experimental configuration also enable the modification or doping with other elements or nanomaterials in the produced carbon nanoparticles, with significant benefits such as being simple in operation, versatile with respect to the selection of other elements or nanomaterials, and scalable in terms of preparing small amounts of samples for research purposes. For example, with the use of ^{13}C-enriched graphite powder or neat ^{13}C powder in the carbon target,

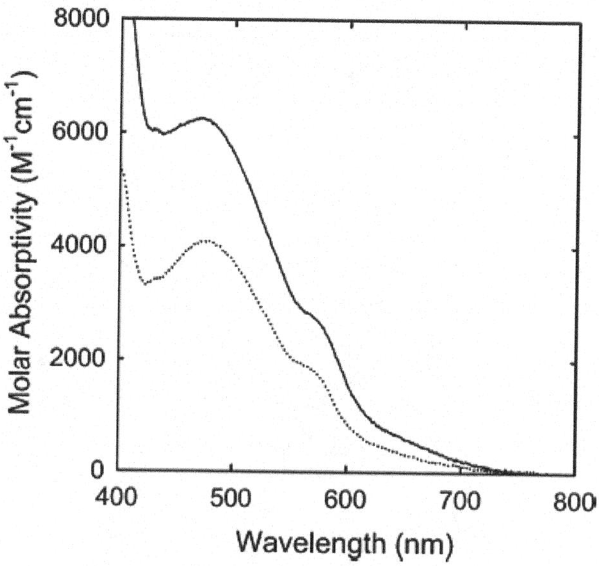

Fig. 2.12 UV/vis absorption spectra of $\mathbf{D_6}$ (Fig. 2.11) in toluene without (——) and with (······) 2 vol% TFAA. (Originally Figure 3 in Li et al. (2005a). Reproduced with permission)

Table 2.4 Absorption parameters of the **D** and **R** compounds (Fig. 2.11) in room-temperature solutions

Compound	Solvent	abs peak (nm)	ϵ_{MAX} (M^{-1} cm^{-1})
R_2	CS_2	495	1850
D_1	Toluene	495	1970
D_2	Toluene	487	2860
D_2	Toluene +2% TFAA	487	1760
$R_{4(e)}$	Toluene	485	2450
$D_{4(e)}$	Toluene	480	3930
$D_{4(e)}$	Toluene +2% TFAA	482	2670
$D_{4(t-3)}$	Toluene	488	2730
$D_{4(t-3)}$	Toluene	470	3990
$D_{4(t-3)}$	Toluene +2% TFAA	486	2450
$D_{4(t-2)}$	Toluene	472	4000
$D_{4(t-2)}$	Toluene +1% TFAA	478	2800
D_6	Toluene	472	6200
D_6	Toluene +2% TFAA	476	4100

Originally Table 1 in Li et al. (2005a). Reproduced with permission

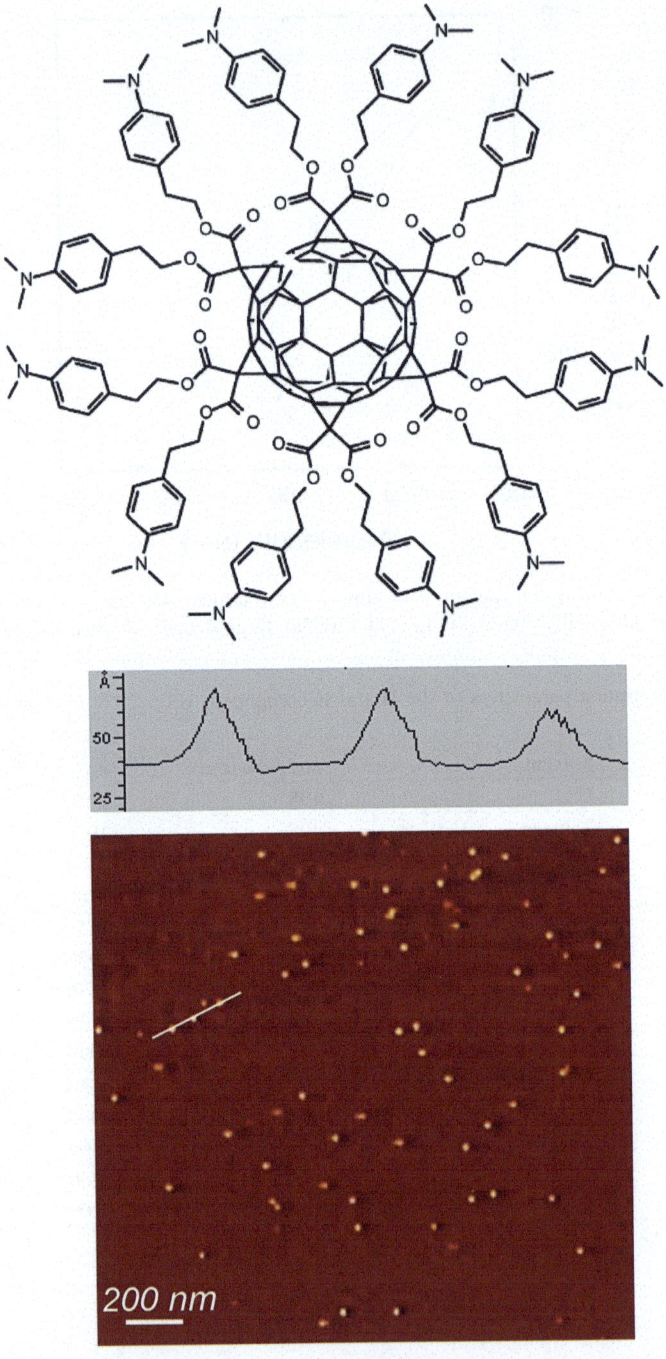

Fig. 2.13 A methano-C_{60} hexakis-adduct with 12 DMA units (upper) (Li et al. 2006) and AFM images of the molecules after treatment with methyl iodide (lower). The sizes for the three highlighted (the height analysis on the top) are around 2.5 nm (25 angstroms) in diameter

Fig. 2.14 A typical laser ablation setup for carbon soot with a substantial content of small carbon nanoparticles. (Sun et al. 2006)

Fig. 2.15 A representative scanning electron microscopy (SEM) image of a carbon soot sample produced by laser ablation

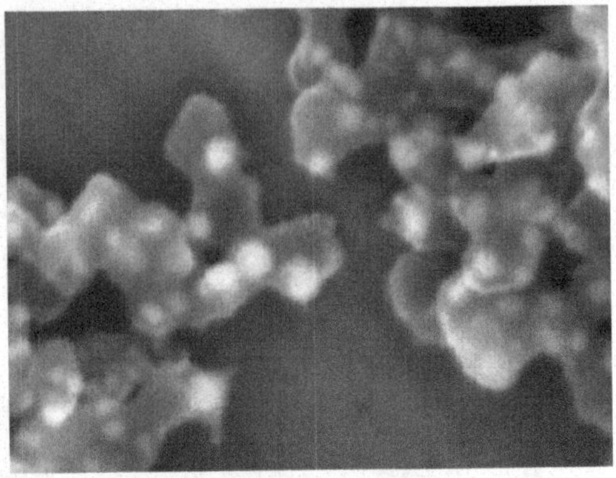

Fig. 2.16 A representative scanning electron microscopy (SEM) image of carbon nanoparticles doped/embedded with nano-nickel

carbon nanoparticles of high ^{13}C contents (or nearly neat ^{13}C nanoparticles with the use of ^{13}C cement) can be produced (Yang et al. 2009c). The same approach can be applied to the doping of heteroatoms like N and S and/or metal elements in the structures of carbon nanoparticles. Experimentally, these elements for doping can be mixed with or embedded in the carbon powders used to make the carbon target for laser ablation. For example, a sample of carbon nanoparticles doped with nickel metal was thus prepared (Fig. 2.16).

Among advantages of laser ablation for carbon nanoparticles are the ability to alter experimental parameters and conditions in favor of smaller particle sizes and the versatility in terms of the modification and doping discussed above. However, the setup for laser ablation is hardly cheap, and the scale-up for production in much larger quantities is generally a significant challenge.

1.2.2 Arc Discharge

Conceptually, arc-discharge method (Fig. 2.17) shares the same strategy with laser ablation in terms of the high-energy input to result in the evaporation of carbon, for which electric arc in the former replaces powerful laser irradiation in the latter. In fact, both were originally used in the preparation of fullerenes and then carbon nanotubes. All of these carbon nanomaterials can be prepared on the same setups, except for the needs of different catalysts and/or processing conditions. The easiest is the preparation of carbon nanoparticles, requiring no catalysts and less stringent experimental conditions, which is likely responsible for the observed consistent presence of carbon nanoparticles as impurities in as-produced fullerene and carbon nanotube samples. The arc-discharge apparatus can be set up relatively inexpensively, with the power unit readily acquired commercially for only several hundred dollars from a welding equipment supplier, much cheaper than a powerful laser source and also with much less maintenance costs. The operation is also generally more efficient than that of laser ablation, capable of producing samples in larger quantities.

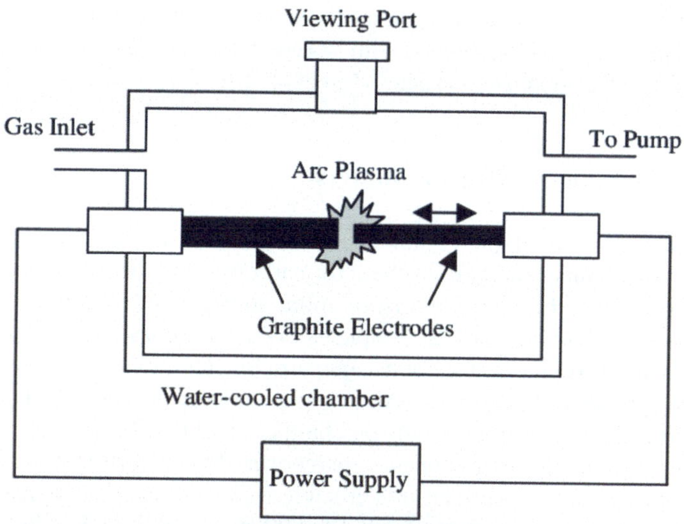

Fig. 2.17 Upper: schematic for a typical arc-discharge setup (From Kingston and Simard (2003)). Lower: a photo showing the inside of the arc-discharge chamber in the setup at Clemson University

In arc discharge, two carbon rods are used as anode and cathode. Since only the rod at the anode side is consumed, that rod may be modified in composition and/or configuration to allow similar doping of ^{13}C, heteroatoms, and/or metal elements in the produced carbon nanoparticles. For example, a hollow carbon rod may be filled with various mixtures that contain the selected elements to be used in the arc discharge (Liu et al. 2014). It is also feasible to prepare the anode rod from scratch, similar to the carbon target for laser ablation discussed above, which allows much higher ^{13}C contents, for example, in the resulting carbon nanoparticles. ^{13}C-enriched or highly enriched carbon nanoparticles are valuable for ^{13}C NMR characterizations

(LeCroy et al. 2014) and in toxicity or mechanistic studies, such as the use of ^{13}C as a quantitative tracer in vivo (coupled with isotope-ratio mass spectroscopy) in toxicity evaluations of carbon nanoparticles (Liu et al. 2014).

1.2.3 Carbonization of Organic Matters

Carbon-rich species or simple organic matters can be used as precursors in thermal processes mostly under inert atmosphere for carbonization into particles of various sizes. Among many different precursors, more advantageous are those already in particle forms, which thus serve as templates in the carbonization to be more favorable toward the formation of carbon nanoparticles. One example was the use of specifically synthesized polymeric nanoparticles, such as polypyrrole nanoparticles prepared with the use of microemulsion (Jang et al. 2002; Wang et al. 2011). The suspended polypyrrole nanoparticles were recovered via centrifugation. Upon drying, the polypyrrole nanoparticles in a crucible were carbonized in a quartz tube in a furnace under inert atmosphere (heated to 800 °C at a rate of 5 °C/min and then held at 800 °C for 5 h with the purging of argon gas) (Wang et al. 2011). The soot-like material thus obtained appeared rather similar to the samples produced by laser ablation and arc discharge, and it was treated the same as the other samples in the further processing for harvesting small carbon nanoparticles. The carbon nanoparticles sourced from originally polypyrrole nanoparticles apparently contained N (in terms of doping, embedding, and/or other configurations). As to be highlighted in the following chapters, the CDots prepared with the polypyrrole-sourced carbon nanoparticles were no different from those similarly prepared with other carbon nanoparticles without any N content in the nanoparticles.

The example highlighted above from the carbonization of specifically synthesized polypyrrole nanoparticles is also meaningful in terms of introducing heteroatoms (N in the case of polypyrrole) into the resulting carbon nanoparticles and subsequently the corresponding CDots. Other elements can be similarly "implanted" in the produced carbon nanoparticles by using the same synthesis strategy with organic precursors containing the selected elements.

The processing conditions used for the carbonization of polypyrrole nanoparticles, including the 800 °C for 5 h in argon atmosphere described above, are aggressive but necessary. As to be discussed in more detail in following chapters, many research groups in their studies of "one-pot carbonization syntheses of carbon dots" reported in the literature grossly underestimated the difficulty in the thermal carbonization of organic matters, thus applied mostly much too mild processing conditions to be sufficient for the desired carbonization, which consequently yielded carbon-deficient or even mostly carbon-less "carbon quantum dots." Described below is an example on how the author learned the requirement for severe conditions in the carbonization of organic precursors.

For education and outreach to pre-college students and the public, the Sun research group at Clemson University designed a science project to demonstrate that a barbeque party could be a part of making CDots of properties suitable for various applications (Wang et al. 2013). Specifically, a small piece of beef patty was used as the starting material for carbon nanoparticles and then CDots. For the carbonization purpose, the beef patty was put on a fire grill as close to the fire as possible and kept there for the duration of the party (about 4 h). In the end, the beef patty did appear charred, though soon it was realized that the char was only superficial. Thus, the surface potion of the charred beef patty was collected and broken into as smaller pieces as possible. The resulting sample had a gray-like color with some black spots. The lack of substantial carbonization as originally anticipated was rather disappointing. Nevertheless, a portion of the sample was further heated in a conventional household oven for many hours, which apparently still did not result in the desired carbonization. By then it was understood that the planned carbonization of the small piece of beef patty into carbon soot-like materials would not happen, unless more aggressive processing conditions were applied. Therefore, another portion of the sample from the barbeque party was washed with acetone, and the cleaned sample after the removal of acetone was placed in the quartz tube in a furnace for carbonization at 800 °C for 5 h, namely, in the same setup and under the same conditions as those for the carbonization of polypyrrole nanoparticles discussed above. This whole process made everyone in the project team feel silly for the initial naive thinking of "barbeque for carbonization."

Many commercially sourced "carbon nanopowder" samples are manufactured by high-temperature carbonization of plant materials in the absence of oxygen. These nanopowders are generally somewhat porous, which may facilitate their "breaking up" in various processing protocols for the harvesting of small carbon nanoparticles that are more useful to the preparation of CDots.

1.2.4 Morphological and Structural Characteristics

Carbon nanoparticles in soot samples produced by laser ablation, arc discharge, and the carbonization of organic matters under commonly used thermal processing conditions (similar to those discussed above, nothing extreme like thermal annealing at 2000+ ^{0}C) are mostly amorphous with some graphitic features in powder X-ray diffraction patterns (obtained with the use of a large amount of sample in the measurement, Fig. 2.18). However, the kind of crystalline carbon nanoparticle structure (Fig. 2.19) imagined by some in the literature is likely imagination only, because there are no reasons in terms of thermodynamic driving forces to predict the formation of such a pretty but strangely ordered particle structure in the rather chaotic environment associated with the production methods discussed above. Also compared in Fig. 2.18 is the X-ray powder diffraction pattern of amorphous carbon nanoparticles from a commercially supplied "carbon nanopowder" sample (produced from the carbonization of plant materials according to the supplier) (Ren et al. 2019).

There may be several ways or angles to look at the largely amorphous carbon nanoparticles. According to the online Wikipedia page (https://en.wikipedia.org/wiki/Amorphous_carbon), in which the IUPAC "Gold Book" *Compendium of Chemical Terminology* and the article by Robertson (Robertson 1986) are

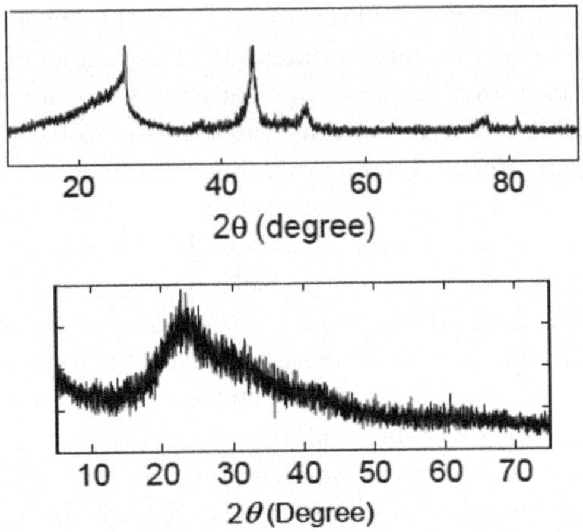

Fig. 2.18 X-ray powder diffraction patterns of carbon nanoparticles harvested from (upper) an arc-discharge sample, as reported in Xu et al. (2011), and (lower) a commercially acquired sample, as reported in Ren et al. (2019)

Fig. 2.19 The structure proposed in some literature reports for individual carbon nanoparticles in "carbon quantum dots" or for "graphene quantum dots"

referenced, "amorphous carbon" (not necessarily at the nanoscale) may be understood as follows: "Amorphous carbon is free, reactive carbon that does not have any crystalline structure." "As with other amorphous solids, some short-range order can be observed." "In mineralogy, amorphous carbon is the name used for coal, soot, carbide-derived carbon, and other impure forms of carbon that are neither graphite nor diamond. In a crystallographic sense, however, the materials are not truly amorphous but rather polycrystalline materials of graphite or diamond within an amorphous carbon matrix." "True amorphous carbon has localized π electrons (as opposed to the aromatic π bonds in graphite), and its bonds form with lengths and distances that are inconsistent with any other allotrope of carbon. It also contains a high concentration of dangling bonds; these cause deviations in interatomic spacing (as measured using diffraction) of more than 5% as well as noticeable variation in bond angle." "The primary method for characterizing amorphous carbon is through the ratio of sp^2 to sp^3 hybridized bonds present in the material. Graphite consists purely of sp^2 hybridized bonds, whereas diamond consists purely of sp^3 hybridized bonds. Materials that are high in sp^3 hybridized bonds are referred to as tetrahedral amorphous carbon, owing to the tetrahedral shape formed by sp^3 hybridized bonds, or as diamond-like carbon (owing to the similarity of many physical properties to those of diamond)."

The above-identified characteristic features of amorphous carbon, especially the lack of any long-range order, should be even more pronounced in amorphous carbon at the nanoscale, which is known to decrease any kind of order in general. Therefore, the observed broad X-ray diffraction patterns of the carbon nanoparticle samples from the production methods discussed above do not necessarily exclude the nanoparticles from their containing small graphitic pieces, rather indicating only that the overall long-range order in these carbon nanoparticles is not sufficient to be reflected in the observed X-ray diffraction patterns. In this regard, the laser ablation production may yield carbon nanoparticles with somewhat more graphitic pieces, because in some of the X-ray powder diffraction results there may be small peak(s) on top of the generally very broad feature. Also according to the quoted statements above, a carbon nanoparticle could be a mixture of three different carbon nanostructures: graphitic pieces, diamond-like pieces, and "random" (amorphous) pieces without any defined bonding forms. In such a morphological picture, the highly ordered particle structure shown in Fig. 2.19 would represent the very extreme for not only the presence of only graphitic pieces but also the requirement for the graphene sheets in such a perfect size/shape distribution to stack and align up perfectly. Again, there are no theoretical and practical justifications to expect the formation of such an imagined strange graphitic structure in any of the available production methods for carbon nanoparticles.

Raman spectra of the carbon nanoparticles produced by different methods are generally comparable, exhibiting the expected "G-band" and "D-band" features. The former is consistent with the presence of graphitic pieces and/or structures, and the latter is associated with defects of different kinds, including those on the nanoparticle surface. Beyond these, however, the Raman results are not so useful to the elucidation of structural details in the carbon nanoparticles. On the other hand, Raman can be used as an efficient tool in the characterization of ^{13}C-enriched carbon nanoparticles, which exploits the quantitative relationship between the Raman shift and the effective atomic weight of carbons involved in the corresponding vibrational mode. The relatively simple Raman-based quantification may be verified by using other techniques, such as isotope-ratio mass spectroscopy measurements.

Commercially supplied carbon nanopowder samples have been used as sources for carbon nanoparticles. The samples from the carbonization of plant materials as precursors, as specified by the suppliers, have been found largely amorphous according to X-ray powder diffraction results (Fig. 2.18). For other commercial samples from unspecified sources (precursors and production methods), there have been wide variations between different suppliers and different batches of a single supplier. For a number of years, the "carbon nanopowder" samples marketed by a major scientific chemical supplier were found to be an excellent source for the harvesting of small carbon nanoparticles, from which high-performance CDots could be prepared. More recently, however, the similarly marketed samples from the same supplier contained mostly graphitic pieces, corresponding to the dominating sharp peaks in observed X-ray powder diffraction patterns. For such samples, there was considerable difficulty in "fishing" out even a small amount of the targeted carbon nanoparticles by repeated applications of the established processing protocol (which was successfully used in the processing of earlier samples from the same supplier). The minor portion thus harvested from the as-supplied sample still exhibited sharp peaks in X-ray powder diffraction patterns, with the substantial content of relatively large multilayer graphitic sheets (up to 100 nm in lateral dimension) confirmed by transmission electron microscopy (TEM) imaging results. Thus, caution should be exercised when sourcing small carbon nanoparticles from commercially acquired samples.

1.3 Properties of Small Carbon Nanoparticles

1.3.1 Solvent Dispersion

The as-produced soot samples populated by carbon nanoparticles of various sizes are generally not dispersible in aqueous or organic media, posing major limitations on their characterizations and uses. The sample morphology is generally such that the carbon nanoparticles are aggregated, as shown in their scanning electron microscopy (SEM) images.

The solvent dispersion to form relatively stable (against precipitation) aqueous or organic suspensions can be achieved by treating the soot samples in processes that would break up the aggregates and/or modify the surface of individual carbon nanoparticles. A relatively simple treatment for such a purpose is the use of oxidative acids, such as nitric acid in different concentrations in experimental protocols that were originally developed and used for the purification of carbon nanotubes. According to available experimental results, the oxidative acid treatment is effective in dispersing the mostly amorphous carbon nanoparticles, but not so much for graphitic nano-pieces. The generally acknowledged effect of the oxidative acid treatment is the oxidation of the carbon nanoparticle surface carbons to form surface-bound oxygen-containing groups, especially carboxylic acids as the terminal stage of the oxidation. Experimentally, the presence of carboxylic acid moieties in the oxidative acid-treated carbon nanoparticles has been supported by results from pH test, FT-IR, and ^{13}C-NMR (based especially on ^{13}C-enriched samples) measurements.

In the literature, the introduction of oxygen-containing groups including carboxylic acids into carbon nanoparticles or other carbon nanomaterials is often considered as a form of "functionalization," namely, that the nitric acid-treated carbon nanoparticles may be considered as "functionalized carbon nanoparticles." Such a consideration is reasonable and in some specific cases can also be very important. However, in most uses of the treated carbon nanoparticles, a critical assessment on the consequence of such functionalization to the properties of the nanoparticles is rather difficult or even impossible, because the assessment itself may require the use of dispersible carbon nanoparticles, for which the oxidative acid-induced functionalization is necessary. Thus, at least on using the carbon nanoparticles for CDots, the dispersible nanoparticles from oxidative acid treatments, often referred to as "pre-processed and selected small carbon nanoparticles," are generally considered as being structurally and property-wise similar to the same nanoparticles while pristine (only in one's imagination because the same pristine nanoparticles are not dispersible), except for the presence of surface oxygen-containing groups. Such a general consideration or assumption has been behind studies of small carbon nanoparticles related to the development of CDots.

Several experimental observations and related practical values are worthy of mentioning. One is on the stability of a carbon nanoparticle suspension against precipitation under gravity or in a centrifuge field depending strongly on the acidity in the suspension, generally much less stable under highly acidic conditions. Such a property is useful to the recovering of the carbon nanoparticles after their oxidative acid treatments. Another is on the use of oxidative acids (including mixed acids such as nitric acid–sulfuric acid mixtures) coupled with vigorous sonication to break up large carbon nanoparticles in the as-produced soot samples into smaller ones. The approach seems more applicable to soot samples from some carbonization methods, in which the nanoparticles are more porous. In the treatment of such samples under the more vigorous processing conditions, smaller carbon nanoparticles could be harvested in high yields due to the intended breaking up of large nanoparticles and probably also the "shaving" of some nanoparticles to make them smaller.

The stable dispersions of carbon nanoparticles have enabled characterizations and studies of optical and other properties by using solution-based techniques.

1.3.2 Optical Properties

Carbon nanoparticles have attracted little attention as optical nanomaterials, much less so than fullerenes, for which their structurally unimpressive or even ugly nature in comparison with the beauty of fullerenes may have played a major role. However, carbon nanoparticles are in fact intrinsically superior in some of the same properties. They have strong optical absorptions over a very broad spectral region (Fig. 2.20), significantly stronger than those of fullerenes in most of the visible spectrum, thus more effective in harvesting solar photons over the same spectral region. More quantitatively, the experimentally determined molar absorptivity values of aqueous dispersed carbon nanoparticles are 50–100 (or roughly 75 ± 25) M_{Carbon}^{-1} cm^{-1} over 400–450 nm, where M_{Carbon} is the carbon molar concentration, namely, the absorptivity measured in terms of the molar concentration of carbon atoms in the carbon nanoparticles. These values compare favorably to the 16 M_{Carbon}^{-1} cm^{-1} of C_{60} at the band maximum in the visible spectrum (around 530 nm).

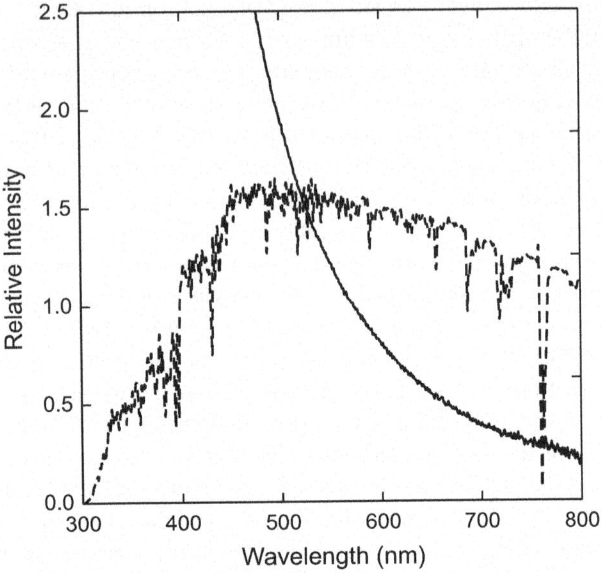

Fig. 2.20 The observed absorption spectrum of broadly distributed CDots in aqueous solution (solid line), which is essentially the same as that of the aqueous dispersed carbon nanoparticles used for the CDots (except for less scattering effects), and the comparison with the solar spectrum at the sea level (dashed line). (Reproduced from Fernando et al. (2015) with permission)

As discussed in the section above, the small carbon nanoparticles harvested from the oxidative acid-treated soot samples can be dispersed in various solvents for stable suspensions. These suspensions are weakly fluorescent, with the observed very broad emission bands covering almost the entire visible spectral region (Fig. 2.21) (Cao et al. 2011b). Due to the presence of carbonyl moieties on the particle surface, their associated n-π* transitions may interfere with the optical absorptions of the carbon nanoparticles at wavelengths of around 350 nm or shorter, so that emission measurements with excitation wavelengths in the near-UV and visible (such as 400 nm to be a favored selection) are more reliable in terms of reflecting the true fluorescence properties of the suspended carbon nanoparticles.

As reported in (Cao et al. 2011b) (Fig. 2.22), fluorescence emission spectra and quantum yields were both clearly solvent dependent, with the spectrum in DMSO particularly narrow in bandwidth, similar to that found in spectra of typical CDots. Correspondingly, the observed fluorescence quantum yield in DMSO was several times of those in other solvents (Table 2.5). As stated for the explanation of the observations (Cao et al. 2011b): "The presence of solvent was apparently necessary for the observation of luminescence emissions from carbon nanoparticles. No luminescence could be detected for the same particles under solvent-free conditions (deposited on solid substrate, for example). This has generally been the case for other naked carbon nanoparticles from different sources and preparations. Thus, it seems logical to assume that the solvent molecules may play the role of passivating the particle surface, similar to but understandably less effective than that of the surface passivation agents in the much more fluorescent functionalized carbon nanoparticles (dubbed "carbon dots")." "In this regard, the higher luminescence quantum yields of the carbon nanoparticles suspended in DMSO might simply be due to

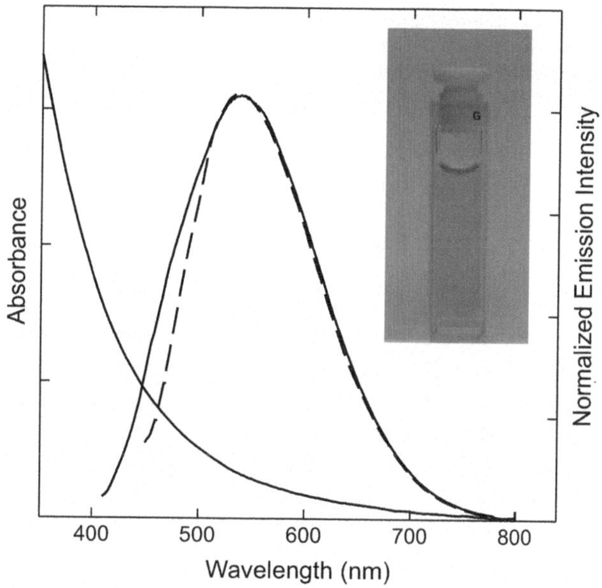

Fig. 2.21 Absorption and emission (400 nm excitation: ——; 440 nm excitation: — —) spectra of the aqueous suspended carbon nanoparticles, with a photo of the suspension in the inset. (Originally Figure 2 in Cao et al. (2011b). Reproduced with permission)

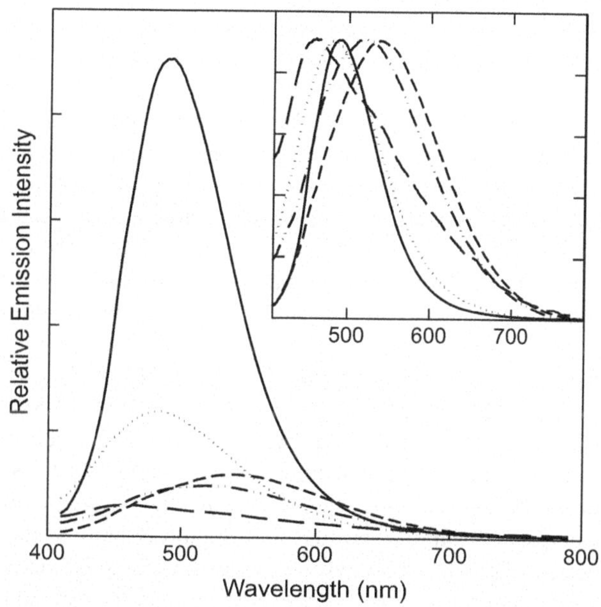

Fig. 2.22 Fluorescence spectra (400 nm excitation, normalized in the inset) of the carbon nanoparticles in DMSO (——), DMF (• • •), water (– – –), ethanol (– • • –), and methanol (— —). (Originally Figure 3 in Cao et al. (2011b). Reproduced with permission)

Table 2.5 Fluorescence quantum yields of the carbon nanoparticles in different solvents

Solvent	$\Phi_F(\%)$ for excitation at 440 nm (400 nm)
DMSO	0.84 (0.73)
DMF	0.29 (0.25)
Water	0.24 (0.16)
Ethanol	0.2 (0.14)
Methanol	0.14 (0.09)

Originally Table 1 in Cao et al. (2011b). Reproduced with permission

relatively more effective particle surface passivation by DMSO molecules (conceptually similar to weak surface noncovalent "functionalization" of the carbon nanoparticles). Therefore, the photoluminescence in all small carbon nanoparticles (naked or functionalized carbon dots) could be sharing the same origin: the emissive energy trapping sites on the passivated surface of carbon nanoparticles" (Cao et al. 2011b). Looking back, the particularly narrow emission bandwidth in DMSO was also consistent with the unusually strong surface passivation effect in this solvent.

Despite the exceptionally strong fluorescence in DMSO, the observed fluorescence quantum yields of the small carbon nanoparticles dispersed in the different solvents are all less than 1% (Table 2.5), but even these low yields are still significantly higher than those of fullerene C_{60} and derivatives. Thus, the carbon nanoparticles are generally considered as weakly emissive, versus fullerenes as essentially nonemissive.

It should be stressed that the carbon nanoparticles in the relatively stable (against precipitation under gravity) solvent dispersions are those harvested from the oxidative acid-treated samples, with some oxygen-containing functionalities such as carboxylic acids on the nanoparticle surface that facilitate the dispersion of the nanoparticles due to improved solvation. Thus, the carbon nanoparticles may be considered as being "functionalized" by the surface moieties and their photoexcited state properties and processes influenced by the salvation. Strictly speaking, in order to understand the intrinsic fluorescence emission properties of carbon nanoparticles, the optical spectroscopy experiments would have to be done on "pristine" carbon nanoparticles. In fact, efforts have been made in the Sun research group at Clemson University to prepare such pristine samples by depositing the processed carbon nanoparticles onto a substrate to ensure their well dispersion on the surface, followed by thermal annealing to remove the oxygen-containing fnctionalities on the nanoparticle surface introduced in the processing of the nanoparticles. So far, however, the results obtained with such sample specimens have shown no detectable fluorescence emissions, suggesting that pristine small carbon nanoparticles are most likely much less fluorescent than their surface-oxidized and solvated counterparts.

The discussion above on nearly pristine carbon nanoparticles is consistent with the notion that surface passivation is required for the observation of significant photoluminescence emissions from the nanoparticles, where the passivation could be as benign as the presence of carboxylic acids or the like on the nanoparticle surface

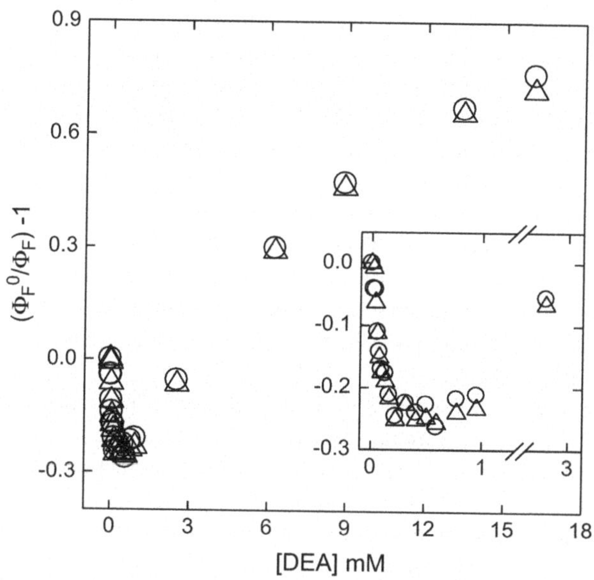

Fig. 2.23 Stern–Volmer plots for the quenching of photoluminescence intensities (400 nm excitation: -o-; 440 nm excitation: -Δ-) of the carbon nanoparticles by diethylamine (DEA) in aqueous solution. (Originally Figure 4 in Cao et al. (2011b). Reproduced with permission)

and/or salvation effects. A more vivid demonstration on the passivation effect was found in the study of photoluminescence emissions from aqueous dispersed carbon nanoparticles with different concentrations of diethylamine (DEA) (Cao et al. 2011b). The initial increases in observed photoluminescence intensities at very low DEA concentrations (Fig. 2.23) could be attributed to adsorption-like interactions between DEA molecules and the nanoparticle surface, with the associated passivation effect enhancing the photoluminescence emissions.

1.3.3 Photoinduced Redox Characteristics

As discussed earlier, fullerenes are famously electron-deficient, with strong charge-transfer interactions with electron donors and, in some cases, the formation of charge-transfer complexes in excited states or even in the ground state. Carbon nanoparticles are probably not so different from other powdery carbon materials, such as carbon black, in terms of being more electron rich than deficient (e.g., Pd/C as a popular catalyst for reduction reactions). In the photoexcited states, however, the carbon nanoparticles dispersed in various solvents can be both electron acceptors and donors, as observed experimentally. As acceptors, photoluminescence emissions of the carbon nanoparticles could be quenched via charge-transfer interactions with electron donors such as amine molecules, similar to what are known for photoexcited fullerenes. The photoexcited carbon nanoparticles are even more potent as electron donors, as reflected in results from studies on the quenching of their photoluminescence emissions by a wide variety of metal cations. Many of the reported quenching constants from Stern–Volmer plots are outrageously large, in some cases orders of magnitude larger than what is possible with diffusion-limited quenching process. These may be explained by a combination of at least two effects. One is that the quenching requires only limited diffusion of the quenchers (here the metal cations), namely, a substantial contribution of near-neighbor static quenching due to adsorption-like pre-association of the metal cations with the nanoparticles being quenched. The other, also due to limited diffusion requirement, is that the electronic sphere of a photoexcited carbon nanoparticle extends significantly beyond the surface of the nanoparticle, thus facilitating the quenching without much diffusion of the metal cations.

The redox interactions responsible for the photoluminescence quenching could also be productive for some metal cations, such as silver, gold, and platinum cations, resulting in reduction for the deposition of the corresponding metal onto the nanoparticle surface (Fig. 2.24) (Xu et al. 2011). For example, the photolysis of aqueous suspended carbon nanoparticles in the presence of silver cation resulted in the surface doping of the nanoparticles with silver metal, corresponding to the

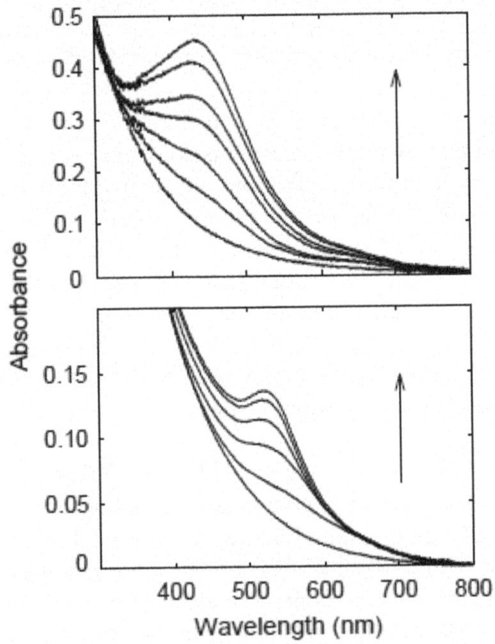

Fig. 2.24 UV/visible absorption spectra (2 mm optical path-length) of the aqueous suspended carbon nanoparticles with progressively increasing levels of Ag^0 coating (upper, with increasing photolysis time in the arrow direction of 0, 1, 2, 3, 4, 6, and 9 h) and similarly Au^0 coating (lower, 0, 1, 2.5, 5, 8, and 10 h). In the control without the presence of carbon nanoparticles, photoirradiation of the aqueous Ag^+ solution under the same experimental conditions resulted in no Ag^0 absorption. (Originally Figure 1 in Xu et al. (2011). Reproduced with permission)

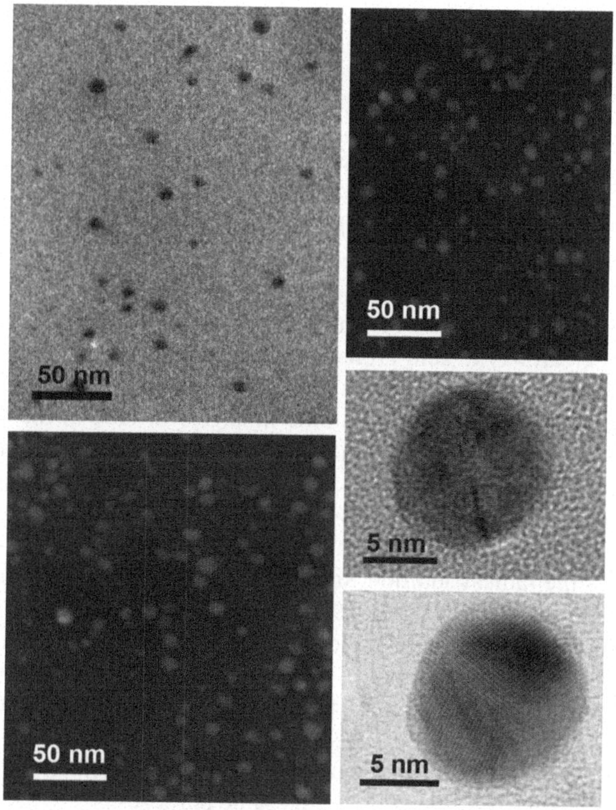

Fig. 2.25 Transmission electron microscopy (TEM) images of the carbon nanoparticles with Ag^0 coating (upper-left: TEM; upper-right: Z-contrast) and similarly with Au^0 coating (lower-left: Z-contrast). The lower-right insets are high-resolution TEM images for the Ag^0-coated (upper) and Au^0-coated (lower) carbon nanoparticles. (Originally Figure 3 in Xu et al. (2011). Reproduced with permission)

progressive increases of the silver plasmon absorption band over time (Fig. 2.24). The silver on the nanoparticle surface could be verified in direct electron microscopy imaging analyses (Fig. 2.25). The same photoreductive surface doping of the carbon nanoparticles by gold and platinum could readily be accomplished in similar experiments, in which the role of photoexcited carbon nanoparticles as potent electron donors is made evident.

The redox characteristics of fullerenes as potent electron acceptors have been explored for applications in optoelectronics and other technological areas. The small carbon nanoparticles with both electron acceptor and donor characteristics in the photoexcited states may find more applications, especially for the fact that the photoinduced redox properties of the nanoparticles are not only preserved but also enhanced upon their effective surface passivation in CDots, as described in more detail in later chapters.

2 Carbon Nanoparticles Versus Fullerenes in Selected Reactions

Fullerenes are known for their addition reactions, yielding methano-fullerene and pyrrolidino-fullerene derivatives, and others such as Diels–Alder adducts, which are related to the electron-deficient nature of fullerene cages. These reactions may not be applicable to the carbon nanoparticles because of different electronic characteristics and bonding patterns. As stressed earlier, unlike fullerenes as stoichiometrically defined molecules, carbon nanoparticles are nanostructured species with surface and other defects, some of which are oxidized in the oxidative acid treatment to become carboxylic acid groups. These functionalities on the nanoparticle surface can be activated for reactions with nucleophiles, especially primary and secondary amine molecules to form amide linkages. In fact, in the derivatization (functionalization) of the carbon nanoparticles, amidation has been particularly popular, where the reaction involves the nanoparticle surface-bound carboxylic acid moieties, not the surface carbon bonds of the nanoparticles.

One reaction that is apparently shared between fullerenes and small carbon nanoparticles is the radical addition, such as in the radical-initiated copolymerization of fullerene cages (Bunker et al. 1995; Cao and Webber 1996; Ford et al. 2000) or carbon nanoparticles with vinyl molecules (Yang et al. 2018b). In fact, the ability for fullerene cages to engage in radical additions has earned C_{60} the famous label as an effective "free radical scavenger," with a wide range of potential applications from uses in beauty products to cancer prevention-therapy pursued in extensive investigations. Then logically, an interesting question is whether the small carbon nanoparticles could be used for some of the same purposes.

C_{60} is soluble in a number of organic solvents, including nonpolar solvents such as toluene, which makes it easier to select a solvent system suitable for both C_{60} and the vinyl molecule. The originally reported copolymerization reactions of C_{60} with styrene were carried in toluene. However, a similar selection of a common solvent system for the carbon nanoparticles and styrene becomes somewhat more challenging due to their different preferences to polar and nonpolar solvents, respectively. In principle strong solvents like DMSO could disperse/dissolve both reactants, but in practice to keep these solvents completely dry for reactions involving free radicals is more difficult than one would expect. These technical issues made solution-phase copolymerization reactions of the small carbon nanoparticles with styrene or methyl methacrylate rather difficult (Yang et al. 2018b). To overcome the technical difficulty, the same reactions could be carried out more successfully in the solid state with vinyl compounds that are solids under ambient conditions, such as *N*-vinylcarbazole (Fig. 2.26) (Yang et al. 2018b).

Experimentally, the small carbon nanoparticles harvested from the nitric acid-treated carbon soot sample were mixed well with *N*-vinylcarbazole, and the solid-state mixture was heated for the thermally induced radical polymerization. The formation of poly(*N*-vinylcarbazole) (PVK) was readily confirmed by results from relatively straightforward characterization experiments, so was the incorporation of

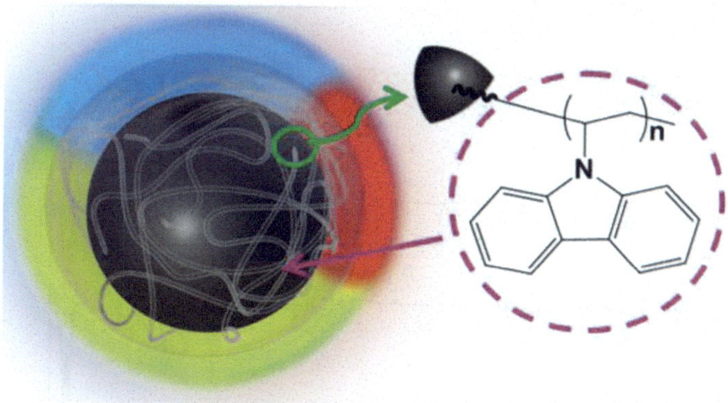

Fig. 2.26 A cartoon illustration on copolymers from thermally initiated radical reactions of *N*-vinylcarbazole with small carbon nanoparticles. (Reproduced from Yang et al. (2018b) with permission)

the carbon nanoparticles in the copolymer. Among the strong evidence for the structural incorporation (instead of simple encapsulation) is the observed substantial change in optical properties after the polymerization reaction, more specifically the change in photoluminescence emission properties. As an example, the observed photoluminescence intensities (quantum yields) of the carbon nanoparticles in the copolymer were much higher than those of the same carbon nanoparticles embedded in PVK films, consistent with the carbon nanoparticle surface functionalization and the associated effective surface passivation for bright and colorful emissions in CDots, thus the designation of PVK-CDots for the copolymer. For comparison, the same reaction conditions were applied to a similar mixture of C_{60} with *N*-vinylcarbazole to obtain PVK-C_{60} copolymer (Fig. 2.27) (Yang et al. 2018b).

The selection of PVK as the primary structure for the copolymers is obviously related to the popular use of PVK in optoelectronic devices for its well-known hole transport function. Similarly popular in these devices, fullerene molecules such as PCBM are used in the active layers, including their serving as photon absorbers. As discussed earlier, small carbon nanoparticles are generally more absorptive than fullerenes over the visible spectrum. This is also true in the comparison between PVK-CDots and PVK-C_{60} copolymers (Fig. 2.28).

The comparison between the two copolymers also raised issue on the possible origin for some of the reported visible fluorescence emissions from derivatized C_{60} samples, which might actually be due to the "contamination" by other functionalized nanoscale carbon species in the samples. In the absorption spectrum of the PVK-C_{60} in dichlorobenzene solution (Fig. 2.29) (Yang et al. 2018b)), the feature around 708 nm is characteristic of mono-functionalized C_{60} cages (commonly referred to as C_{60} mono-adducts), suggesting the presence of such structural elements in the copolymer. As discussed earlier, the parent and derivatized C_{60} cages are generally weak in fluorescence emissions and even weaker for multi-adducts including copolymers with incorporated C_{60} cages. Similarly in (Yang et al. 2018b),

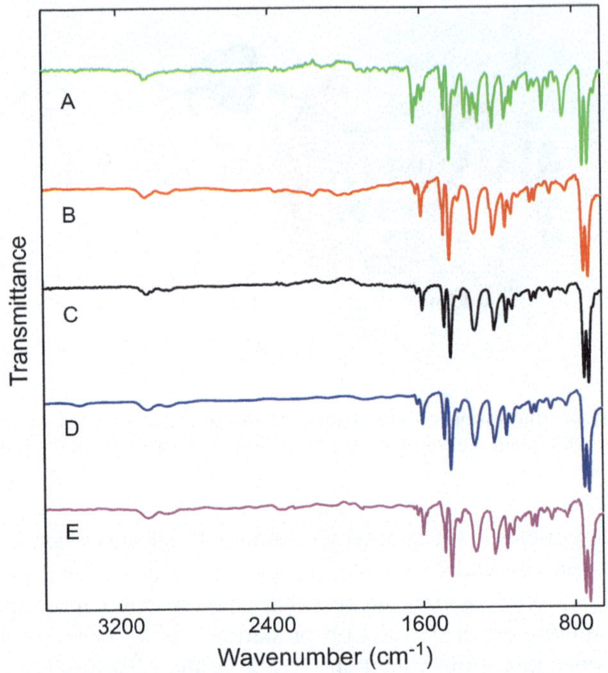

Fig. 2.27 FT-IR spectra of *N*-vinylcarbazole (A), PVK-CDots (B), PVK-C_{60} (C), and PVK polymers from the control experiment (D) and acquired commercially (E). The spectra appear very similar because they are all dominated by PVK polymers, suggesting no other species of significant IR signals formed in the copolymerization reactions. (Originally Figure 5 in Yang et al. (2018b). Reproduced with permission)

the PVK-C_{60} was also found to be only weakly emissive, with observed fluorescence quantum yields generally less than 0.1% for excitation wavelengths across the visible spectral region. In fact, because the emissions due to the functionalized C_{60} cages were so weak, the observed excitation wavelength-dependent fluorescence spectra of the PVK-C_{60} (Fig. 2.29) were attributed to some minor presence of CDots-like species in the sample formed under the thermal processing conditions for the copolymerization reaction. The distinct emission features around 725 nm in the observed fluorescence spectra (Fig. 2.29), characteristic of C_{60} mono-adducts, were considered as a more reliable signature for the emissive properties of the PVK-C_{60} (Yang et al. 2018b).

The comparison between fullerenes and carbon nanoparticles in radical reactions and for their incorporation into copolymers is interesting for both fundamental understanding and technological application purposes. In an effort on reconciling the optical properties between fullerenes and the small carbon nanoparticles, a graduate student in the Sun research group at Clemson University did a project nicknamed "beauty and beast," with fullerenes as the beauty and their derived nanoscale carbon species as the beast. Experimentally, C_{60} molecules and small clusters

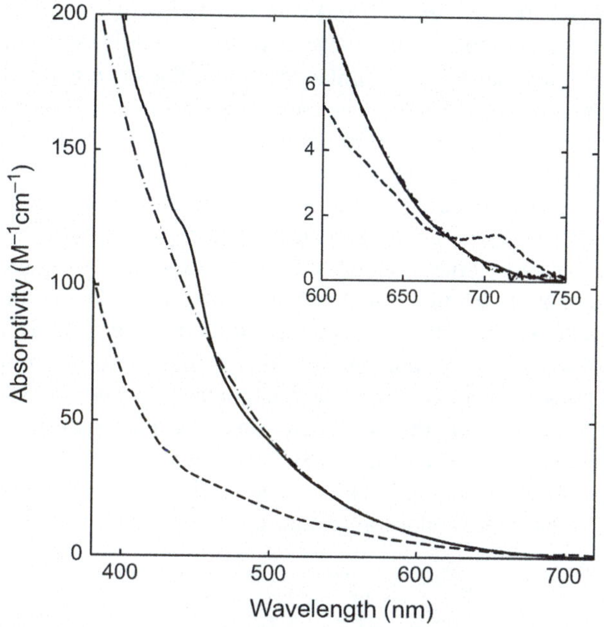

Fig. 2.28 Absorption spectra (enlarged for the 600–750 nm spectral region in the inset) of the PVK-CDots in DMF solution (——) and PVK-C_{60} in dichlorobenzene solution (- - - -), both in the absorptivity unit reduced to per mole of carbon and also the aqueous suspended small carbon nanoparticles (-·-·-). (Originally Figure 2 in Yang et al. (2018b). Reproduced with permission)

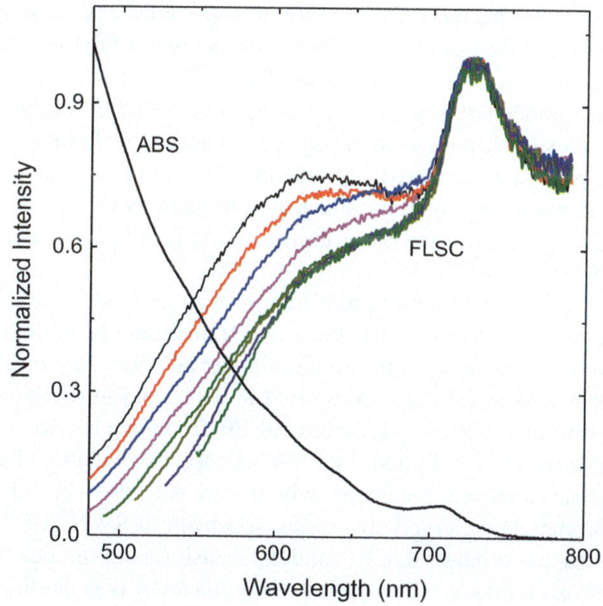

Fig. 2.29 Absorption (ABS) and fluorescence (FLSC, from left to right corresponding to excitations at 400 nm, 420 nm, 440 nm, 460 nm, 480 nm, 500 nm, 520 nm, and 540 nm) spectra of the PVK-C_{60} in dichlorobenzene solution. (Originally Figure 6 in Yang et al. (2018b). Reproduced with permission)

(Sun and Bunker 1993; Sun et al. 1995) were reacted with amino molecules under chemical functionalization or thermal carbonization conditions, in comparison with the small carbon nanoparticles in similar reactions. The chemically derivatized C_{60} cages, similar to many other C_{60} mono- and multi-adducts, remained only weakly fluorescent. However, for the samples obtained from the "functionalization" of C_{60} with amino molecules under thermal carbonization conditions, in which the functionalized species were structurally more like C_{60} monomer or small clusters "covered" by nanoscale carbon domains attached with amino molecules, the fluorescence emissions became more comparable with those of the pre-processed and selected small carbon nanoparticles functionalized by similar amino molecules.

By extrapolation, one might expect that fullerene cages could be crushed to become disordered and more nanoparticle-like, with their surface functionalization in various schemes (as discussed in more detail in later chapters) to yield CDots-like structural configurations and optical absorption and emission properties.

In a more recent study (Ren et al. 2019), the microwave-assisted thermal reaction for the functionalization of small carbon nanoparticles with *N*-ethylcarbazole was also attributed to be radical addition in nature.

3 Carbon Nanoparticles Versus Carbon Nanotubes and Graphenes

There have been many comparisons made between carbon nanotubes and then graphene nanosheets with fullerenes for their shared properties and differences. Similar comparisons with small carbon nanoparticles have largely been ignored, despite the fact that they are more in common in some of the structural and electronic characteristics. It may also be argued that the properties of carbon nanoparticles could help the understanding of some behaviors found in carbon nanotubes and graphene nanosheets, such as those more relevant to this book concerning the optical properties of functionalized carbon nanotubes and graphene nanosheets. These properties are associated with the electronic transitions in the one- and two-dimensional carbon allotropes and also the inevitable presence of structural and other defects in these carbon nanomaterials.

Carbon nanotubes are different in some fundamental ways from fullerene molecules in electronic structures and transitions. In both single-walled (SWNTs) and multiple-walled carbon nanotubes (MWNTs), the optical absorptions are due to π-plasmons (simply understood in some literature reports as the "π-electronic collective excitations") (Rance et al. 2010), which are not available in individual fullerene molecules. In addition, SWNTs are also capable of absorptions associated with their electronic bandgaps, which vary with the nanotube diameters and chiralities, so that an observed absorption spectrum of SWNTs is typically a superposition of structured bands due to bandgap transitions in the near-IR on top of the extremely broad π-plasmon absorption. Only the latter is generally observed in the absorption spectrum of MWNTs (Fig. 2.30).

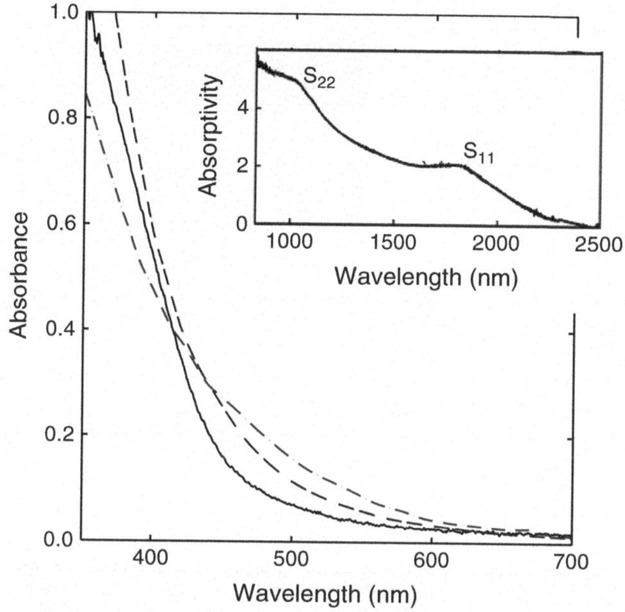

Fig. 2.30 Absorption spectra of SWNTs (-·-·-, from arc-discharge production, inset: the bandgap absorption bands in the near-IR), MWNTs (— —), and carbon nanoparticles (——)

Optical absorptions of graphene nanosheets in the visible spectrum are apparently due to π-plasmons as well, similar to those in carbon nanotubes. Shown in Fig. 2.31 is an observed absorption spectrum of few-layer graphene nanosheets in a solvent dispersion.

Experimentally, the small carbon nanoparticles exhibit similarly broad optical absorption features, both in solvent dispersion and in polymer matrix, and the observed absorption spectra are in general agreement with those of carbon nanotubes (Fig. 2.30) and graphene nanosheets (Fig. 2.31). Mechanistically, the carbon nanoparticles obviously lack of any long-range π-electronic configurations, so that their π-plasmons associated optical properties may have to be viewed somewhat differently. Instead of the π-electron conjugations in one-dimension in carbon nanotubes and two-dimension in graphene nanosheets (including the "interlayer coulomb interactions" as an explanation in the literature on multiple-layer graphene nanosheets), in the small carbon nanoparticles, the π-plasmons could be a result of abundant π-electrons confined in the pseudo-zero-dimension. Regardless of the mechanistic details, however, an experimental observation-derived conclusion must be that the small carbon nanoparticles share rather similar optical absorption characteristics with carbon nanotubes (except for the bandgap absorptions in SWNTs) and few-layer graphene nanosheets over the near-UV, visible, and near-IR spectral regions.

Again different from fullerene molecules, SWNTs may have defects on the wall and/or in the ends, and MWNTs (often viewed as concentrically arranged multiple SWNTs) may also contain other defects. Even a "perfect" single-layer graphene

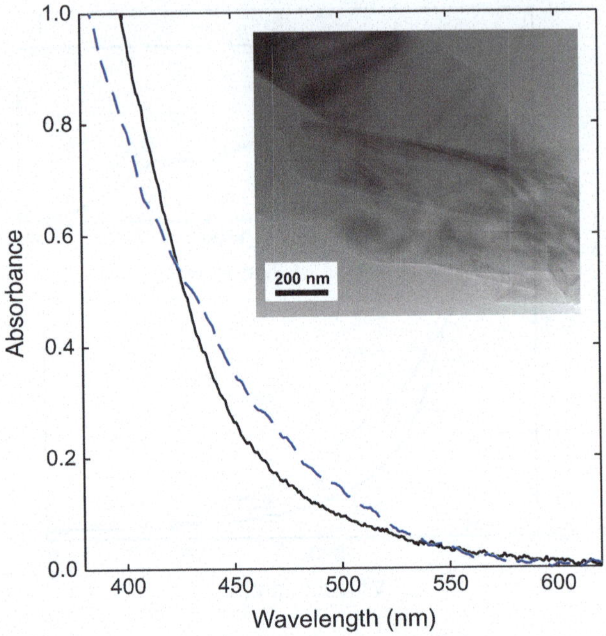

Fig. 2.31 The absorption spectrum (— —) of aqueous dispersed graphene nanosheets (a representative TEM image in the inset) harvested from the oxidative acid treatment (similar to that developed and used for carbon nanoparticles) of a commercially acquired carbon "nanoparticle" sample with dominating content of graphene pieces. The spectrum of aqueous dispersed small carbon nanoparticles (——) is also shown for comparison

sheet would have defective edges or even more defects when the sheet is selectively saturated to create sp^2 islands for the specific version of "graphene quantum dots" (Cao et al. 2013). Effects of the various defects on the electronic and other properties of carbon nanotubes have been investigated in the literature, and also to some extent for graphene nanosheets, but not so much in terms of comparing/referencing such defects with/to those in small carbon nanoparticles (Figs. 1.3). Nevertheless, there are significant experimental observations suggesting some shared characteristics between defects in these different nanoscale carbon allotropes, including defect-derived major effects on optical and photoinduced redox properties.

In carbon nanotubes, holes on the tube wall with dangling bonds are often cited or observed as representative defects. Understandably, similar defects are found in the two-dimensional structures of graphene nanosheets (Fig. 1.3). For nanosheets produced from most syntheses, there are also abundant edge defects, with those in multiple-layer graphene nanosheets similar to the surface defects in carbon nanoparticles (Veca et al. 2009).

"Pristine" carbon nanotubes and graphene nanosheets are not solvent dispersible, making most solution–/dispersion-based experiments virtually impossible. Similar to carbon nanoparticles, however, these carbon nanomaterials can be processed by oxidative acid treatments or the like, in which the carbons at defect sites are oxidized for the introduction of oxygen-containing species, thus enabling their solvent dispersion. In fact, most of the optical absorption spectra in the literature were corresponding to the treated samples, including those shown in Figs. 2.30 and 2.31. One may argue that the spectroscopic and other results obtained with such processed samples do not reflect the true/intrinsic properties of the carbon nanotubes and graphene nanosheets in their pristine forms, despite the fact that the truly pristine is next to impossible, let alone in sufficient quantities of these carbon nanomaterials required for the overwhelming majority of their targeted applications. For example, a sample of SWNTs is almost always contaminated by metal catalysts used in the synthesis, whose removal is often through oxidative acid treatments, which create defects. While it is possible to use high temperatures (>2000 °C) to evaporate the catalyst metals and at the same time "repair" defective nanotubes under the ultrahigh-temperature thermal annealing conditions, the outcome is a mixture of both good news and bad news. The good news is that the nanotube samples from such an extreme treatment can indeed be highly pure in carbon content and also defect-free, with a beautifully shiny metallic appearance. Equally beautiful in terms of clearly defect-free are their Raman spectra and microscopy images. Unfortunately the bad news is really bad and prohibitive, with unintended consequences of the extreme treatment including the new impurities from the conversion of some SWNTs into MWNT-like species made possible by the ultrahigh-temperature conditions. Other than some solid-state measurements, these seemingly beautiful samples cannot be further processed for their intended uses without painstakingly reintroducing defects. Therefore, a takeaway is that the presence of some structural and edge defects in carbon nanomaterials may represent an unavoidable reality. The "pseudo-pristine" samples, defined loosely as those with defects introduced in the production and/or normal best-possible purification and processing protocols, have been the ones offering most realistic opportunities in the development and implementation of the many technological applications. Beyond such a reality, defects in the carbon nanomaterials may also be exploited for unique properties, as made evident in the discovery and development of CDots and their derived/related nanostructures and nanomaterials.

4 Concluding Remarks

Carbon nanomaterials have been at the center of the rapid advancing and expanding research field of nanoscience and nanotechnology. While the excitements about the discoveries of fullerenes, carbon nanotubes, and then graphenes were well justified, the nearly complete ignoring of carbon nanoparticles for their role in carbon nanomaterials was really unfortunate. It is a lesson that "ugly," "disordering," and/or "defective" could actually be highly valuable attributes of nanomaterials, because materials at the nanoscale are intrinsically "imperfect" (except for fullerenes, as one might argue, but they are nanometer-sized molecules). The takeaways:

(1) small carbon nanoparticles offer unique and/or advantageous properties, deserving the title of carbon nanomaterials at the zero-dimension; (2) the nanoparticles share significant structural and other attributes with carbon nanotubes and graphene nanosheets, especially with respect to the defects and their exploitations; (3) the nanoparticles are size-wise closer to molecules (thus more readily dispersed, processed, functionalized/modified, etc.), competing effectively with not only fullerenes but also other macromolecules; and (4) the nanoparticles are benign and nontoxic and easily sourced from abundant and renewable precursors at low to ultralow costs, so that their extremely broad and innovative uses across many technical fields and in new or improved consumer products may be envisaged.

Chapter 3
Carbon Dots: The Beginning

Carbon dots (CDots, Fig. 1.1) are generally defined as small carbon nanoparticles with various surface passivation schemes. As defined, they are structurally built upon the carbon nanoparticles, which as discussed in previous chapters represent a distinct carbon allotrope at the zero-dimension. Historically in the evolution of carbon nanomaterials research and development, however, carbon nanoparticles were often lumped in with "carbonaceous impurities," unwanted and to be eliminated. It was in such a backdrop that CDots were found and initially validated.

1 Leading to the Discovery

1.1 Functionalized Carbon Nanotubes

A number of years before the first report on CDots, bright and colorful photoluminescence emissions were first found in functionalized carbon nanotubes (Fig. 3.1) by the Sun research group at Clemson University. In the work reported in 2000 (Riggs et al. 2000), single-walled and multiple-walled carbon nanotubes (SWNTs and MWNTs, respectively) were processed in a protocol involving oxidative acid (nitric acid) treatment for several purposes. The first was for purification under the assumption that carbonaceous impurities in the nanotube samples should be more readily oxidized than the nanotubes to form oxygen-containing surface moieties, thus significantly more dispersible in the aqueous medium to be washed away. The assumption has largely been validated experimentally over time. The second was to oxidize the defect sites on the nanotubes to impart similar oxygen-containing moieties such as carboxylic acids to be used for chemical functionalization with selected organic molecules. As related, the third was to cut the nanotubes to increase the population of defects at both ends of a shortened nanotube. The processed and cleaned samples of SWNTs and MWNTs were functionalized with oligomers and polymers bearing amino groups in classical amidation reactions. The functionalized

© Springer Nature Switzerland AG 2020
Y.-P. Sun, *Carbon Dots*, https://doi.org/10.1007/978-3-030-41184-8_3

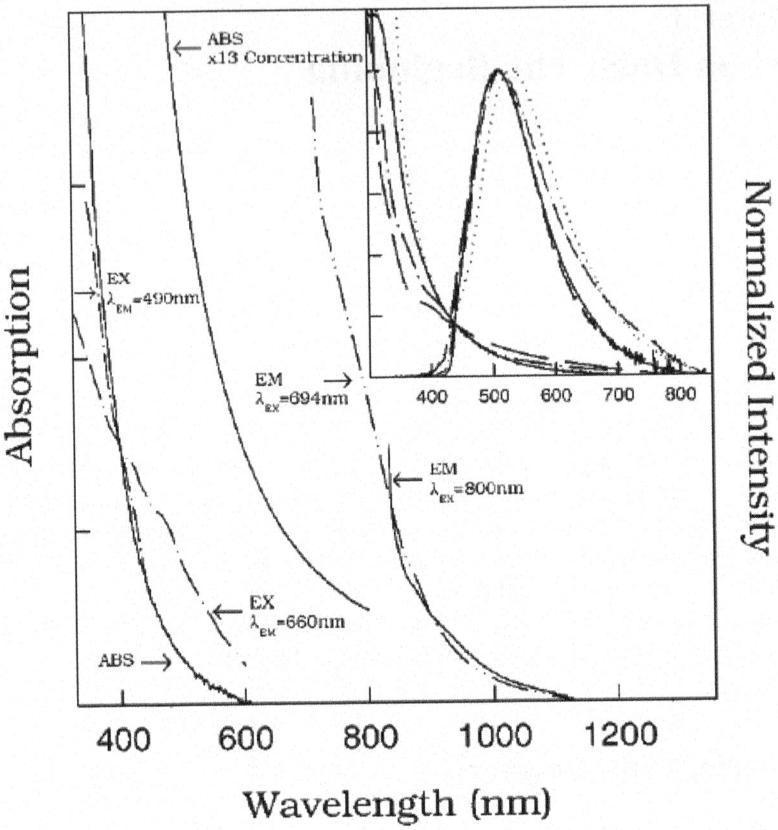

Fig. 3.1 Absorption (ABS), luminescence (EM), and luminescence excitation (EX) spectra of the PPEI-EI polymer-functionalized MWNTs in room-temperature chloroform. Insct: a comparison of absorption and luminescence (440 nm excitation) spectra of the different functionalized MWNTs and SWNTs in homogeneous chloroform solutions at room temperature. (Originally Figure 2 in Riggs et al. (2000). Reproduced with permission)

carbon nanotubes became soluble in common organic solvents and also water when aqueous soluble molecules were used for the functionalization. The resulting solutions were visibly shiny, suggesting photoluminescence emissions in the visible spectral region. The bright and colorful emissions were confirmed in subsequent spectroscopy measurements. The observed emission quantum yields were up to 11% for excitations in the visible spectrum (Riggs et al. 2000), very impressive and surprising, and rather unexpected within the "common sense" of molecular spectroscopy on materials like the carbon nanotubes, which are large in size and extensive in π conjugation. At the time, the bandgap transitions in SWNTs associated with the van Hove singularities were predicted to be deep into the near-IR.

The results described above were several years before the 2002 report in *Science* by researchers at Rice University on the observation of bandgap fluorescence emissions from small-diameter "pristine" (not treated by oxidative acid or the like) semiconducting SWNTs that were well dispersed in solution. Nevertheless, it was clear to the Sun research group at Clemson University from the very beginning that the "green emissions" (dubbed for the green emission color to naked eyes) from the functionalized carbon nanotubes could not be bandgap in nature and instead must be associated with some structural features that were shared by both SWNTs and MWNTs. They proposed in the initial publication (Riggs et al. 2000) that such shared features would have to be the functionalized defect sites on the nanotubes. Several years later in the work on the bandgap fluorescence of SWNTs, the Rice University researchers found that to keep the nanotubes pristine was critical to the observation of the fluorescence because defects created by oxidative acid treatment of the nanotubes and/or the nanotube surface functionalization chemistry could readily quench to the degree of diminishing the bandgap fluorescence emissions. Their finding essentially validated the proposed origins of the green emissions, later designated as "defect-derived emissions of carbon nanotubes," for which the passivation of defects through chemical functionalization was more of an accident in the initial study, but then understood as being necessary in subsequent investigations. Also understood subsequently was the inverse correlation between the bandgap fluorescence and the green emissions in SWNTs, such that the former required the nanotubes to be pristine or otherwise diminished due to the quenching by defects and/or surface modification, thus eliminating the latter, while the latter required passivated/functionalized defects, thus diminishing the former.

In terms of nomenclature for the green emissions in carbon nanotubes, "luminescence" or "photoluminescence" was used generally as a more precise and yet rather diffuse description of the defect-derived emissions for a lack of clear understanding of their associated electronic transitions. Strictly, fluorescence is a term used exclusively in photophysics for emissions associated with the starting and ending states of the same multiplicity. Probably for the absence of any evidence for the green emissions being due to electronic states of different multiplicities, "fluorescence" and "luminescence/photoluminescence" have been used loosely and often interchangeably in the literature in the description of these experimentally real and validated but still often considered as being strange emission properties of carbon nanotubes. Similarly in this book, these terms (photoluminescence, luminescence, and fluorescence) are also used loosely and interchangeably throughout.

There was some skepticism in the literature following the initial report on the green emissions in functionalized SWNTs and MWNTs, and the skepticism was associated mostly with concerns on possible artifacts due to carbonaceous impurities in the nanotube samples. In fact, the same concerns and related technical issues were carefully considered and addressed experimentally in the originally reported study, and the conclusion was that while some contributions to the observed green emissions by other minor carbonaceous entities in the samples could not be ruled out completely, the emissions must be due substantially to the functionalized carbon nanotubes. Strongly supporting such a conclusion was the observed high emission quantum yield of up to 11% for the excitation at 400 nm (Table 3.1), which was determined by the relative method in reference to established fluorescence standards (e.g., the quantitatively fluorescent 9,10-bis(phenylethynyl)anthracene).

Table 3.1 Luminescence parameters of the polymer-functionalized carbon nanotubes in solution

Sample	Solvent	λ_{EX} (nm)	Φ^a	τ_1 (ns)	τ_2 (ns)	a_1/a_2
S-MWNT-PPEI-EI	$CHCl_3$	632		2.2	5.6	1.2
		400	0.11	2.3	8.0	2.9
	Water	400	>0.03	2.3	8.9	4
S-SWNT-PPEI-EI	$CHCl_3$	632		1.9	5.8	1.2
		440	0.06			
SWNT-PPEI-EI	$CHCl_3$	400	> 0.003	1.5	7.3	4.2
	Water	365		2.1	9.8	2.3
S-MWNT-PVA-VA	$CHCl_3$	365	>0.03	1.6	6.0	7.3

[a] 9,10-Bis(phenylethynyl)anthracene as the quantum yield standard
From Riggs et al. (2000)

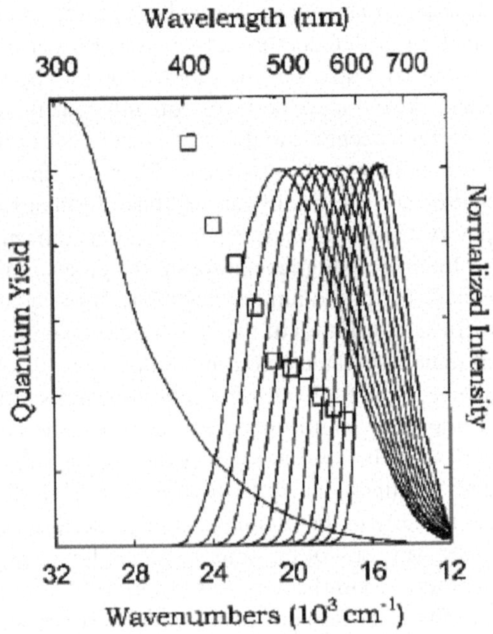

Fig. 3.2 Progressive changes of the luminescence spectra and quantum yields (□) of PPEI-EI polymer-functionalized MWNTs in homogeneous chloroform solution with excitation wavelengths (from left to right, 400–600 nm for the spectra and 400–580 for the quantum yields in 20 nm increment). (Originally Figure 3 in Riggs et al. (2000). Reproduced with permission)

At that wavelength the nanotubes are strongly absorptive, so that they, not the possible carbonaceous impurities, must be dominating the photoexcitation. If only the impurities were emissive, these impurities would have to have an average emission quantum yield of more than 100%, obviously impossible within any existing theoretical frameworks.

There were some other characteristic emission behaviors of carbon nanotubes found in the originally reported study (Riggs et al. 2000). Both the emission spectra and quantum yields determined experimentally were excitation wavelength dependent in a rather characteristic pattern. For the spectra (Fig. 3.2), the band maximum progressively shifted to the red with increasing excitation wavelength, which was accompanied by the "beautifully consistent" gradual narrowing in the bandwidth.

Equally consistent was the gradual but significant decrease in the observed emission quantum yield with the increasing excitation wavelength. These were attributed to the presence of multiple emissive states or species (such as distributions of different defect sites on the nanotubes), as also consistent with the observed non-exponential emission decays and also different decays at different excitation wavelengths (Table 3.1).

A few years after the original work reported in 2000, there were some publications by other research groups on the observation of fluorescence emissions from carbon nanotubes in the near-UV and visible spectral regions, with results generally consistent with those described above. In those reported studies, there were understandably desires and/or attempts to associate the emissions with electronic bandgaps, as those studies were carried out either before the 2002 *Science* paper by the Rice University researchers on semiconducting SWNTs or concerning only metallic SWNTs. Looking back, however, one would conclude that the results from those studies could be better understood in the framework of "defect-derived emissions."

In a work reported in 2002 (Banerjee and Wong 2002), Stanislaus Wong's research group functionalized SWNTs with Wilkinson's catalyst and investigated optical absorption and emission properties of the functionalized nanotubes, in reference to the defect-derived photoluminescence results (Riggs et al. 2000) discussed above. Their luminescence studies in DMSO "indicated that the emission spectrum is strongly dependent on excitation wavelength, which is indicative of the presence of a large number of emitters and absorbers" (Banerjee and Wong 2002). In their discussion of the results, "it seems likely that functionalization facilitates the manifestation of the intrinsic luminescence, emanating from these tubes, through dispersion of these nanotubes as well as trapping of the excitation energy on the nanotube surface itself. The slow motion of the tubes in solution suggests that this energy is not lost rapidly, and thus, the observed quenching and corresponding deactivation rate through molecular motion is slower" (Banerjee and Wong 2002).

Also in 2002, Nagasawa et al. reported the observation of "strong visible emission" under laser illumination of SWNTs hosted in "$AlPO_4$-5 (AFI) crystal" at room temperature, without referencing or comparing with any solution-phase results already available in the literature at the time (Nagasawa et al. 2002). The AFI "host" might be somewhat similar to a polymeric matrix, providing relatively weak functionalization-like effects. They also found that the observed visible emission was highly polarized with respect to the use of linearly polarized excitation light.

Almost at the same time, the Sun research group at Clemson University investigated the luminescence anisotropy of functionalized SWNTs and MWNTs both in solution and in polymeric thin-film matrix (Sun et al. 2002a). As reported, the luminescence was found as being strongly polarized, with significantly positive anisotropy observable even in room-temperature solution. For the functionalization, a hydroxyl-terminated dendron (I_{PEG}, Fig. 3.3) was employed for attachment to the oxidative acid-treated carbon nanotubes via ester linkages. In spectroscopy experiments, luminescence polarization measurements were conducted by using two dichroic sheet polarizers in a right-angle geometry. The polarization factor P and anisotropy r were calculated in terms of the following equations (Sun et al. 2002a):

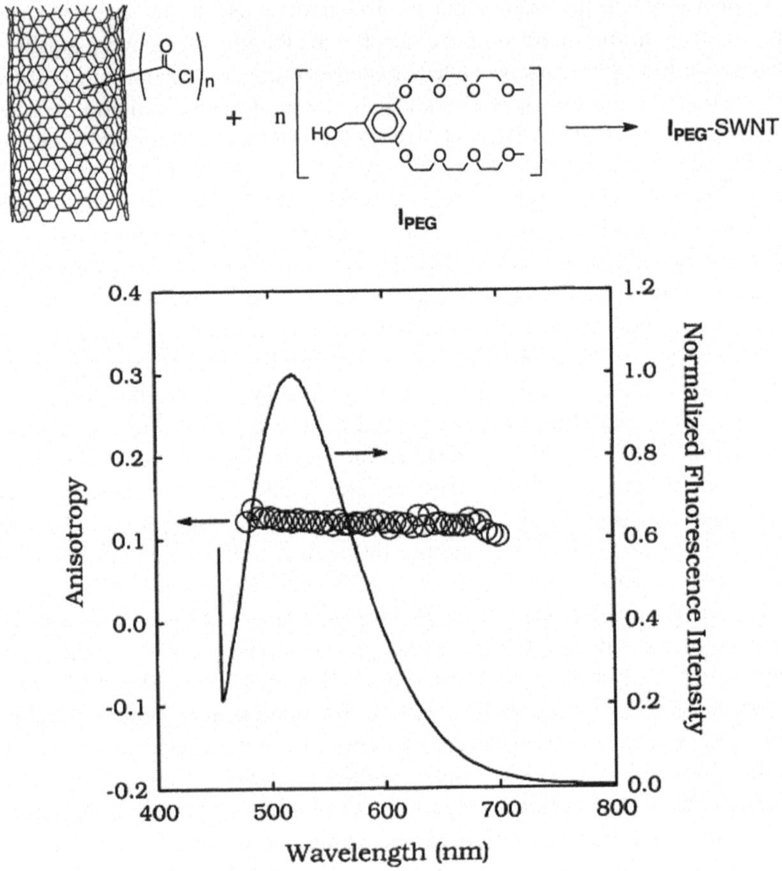

Fig. 3.3 Upper: the functionalization of defects in SWNTs by I_{PEG}. Lower: luminescence anisotropy values at different emission wavelengths for I_{PEG}-MWNT in room-temperature chloroform solution with 450 nm excitation. The luminescence spectrum at the same excitation wavelength is also shown for comparison. (Originally Figure 2 in Sun et al. (2002a). Reproduced with permission)

$$P = \left(I_{\mathrm{HH}} I_{VV} - I_{HV} I_{VH}\right) / \left(I_{HH} I_{VV} + I_{HV} I_{VH}\right), \tag{3.1a}$$

$$r = 2P / \left(3 - P\right), \tag{3.1b}$$

where I_{HH} denotes the luminescence intensities for both excitation and emission polarizers in horizontal positions; I_{HV} for excitation polarizer in horizontal and emission polarizer in vertical positions; and so on. The known theoretical limits are $-0.2 \leq r \leq 0.4$. For I_{PEG}-MWNT in chloroform solution as an example, the r values were essentially independent of emission wavelengths, with an average of ~0.12 at 450 nm excitation (Fig. 3.3). At other excitation wavelengths, the observed r values were similarly emission wavelength independent. However, the luminescence polarization was strongly dependent on excitation wavelengths (Fig. 3.4).

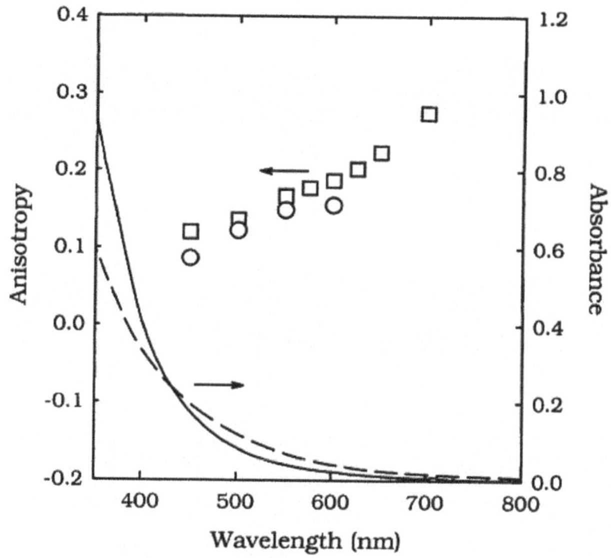

Fig. 3.4 Excitation wavelength dependencies of luminescence anisotropy for I_{PEG}-SWNT (○) and I_{PEG}-MWNT (□) in room-temperature chloroform solutions. The absorption spectra of I_{PEG}-SWNT (- - -) and I_{PEG}-MWNT (—) are also shown for comparison. (Originally Figure 3 in Sun et al. (2002a). Reproduced with permission)

The significantly positive r values (close to 0.3 for I_{PEG}-MWNT at 700 nm excitation, Fig. 3.4) were attributed to an alignment of the emission dipole moment (within the luminescence lifetime) with the absorption dipole moment, but at the same time, the r values were still far from the theoretical limit 0.4.

The same luminescence anisotropy measurements were carried out for the I_{PEG}-SWNT sample in solid poly(vinyl alcohol) (PVA) thin-film matrix. The resulting r values were close to 0.4, essentially at the limit for collinear absorption and emission dipole moments, suggesting that the results in room-temperature solution were due to molecular motion-induced depolarization processes in the photoexcited states of the functionalized carbon nanotubes (Sun et al. 2002a). These findings in the work reported in 2002 were revisited and confirmed (Table 3.2 and Fig. 3.5) in an expanded study of PPEI-EI-SWNT [PPEI-EI = poly(propionylethylenimine-*co*-ethylenimine)] reported in 2006 (Zhou et al. 2006). In that study, the PVA films were stretched to align the embedded nanotubes for absorption polarization measurements. The results (Fig. 3.6) conformed the general understanding that the observed optical absorptions of SWNTs are associated with electronic transitions along the nanotube axis.

The polarization results for both optical absorptions and photoluminescence emissions of the functionalized carbon nanotubes have been puzzling with respect to their mechanistic implications. Because of the defect-derived nature of the underlying photoluminescence emissions, the nearly perfect collinear alignment of the corresponding emission dipole (at least at the longer excitation wavelengths, Table 3.2) with that of the optical absorption along the nanotube axis was surprising

Table 3.2 Luminescence anisotropy (r) values at different excitation wavelengths

Excitation wavelength (nm)	400	425	450	475	500	525	550	575	600
$r_{solution}$ [a]	0.052	0.069	0.10	0.12	0.14	0.15	0.16	0.18	0.19
r_{film}	0.32	0.34	0.35	0.35	0.37	0.38	0.39	0.39	0.39

[a] From repeating the experiments reported in Sun et al. (2002a)
From Zhou et al. (2006)

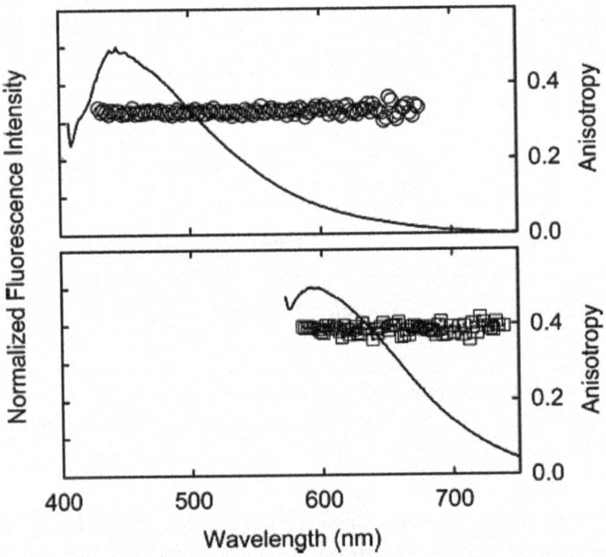

Fig. 3.5 Luminescence emission spectra and corresponding anisotropy values of PPEI-EI polymer-functionalized SWNTs in poly(vinyl alcohol) (PVA) film with the excitation at 400 nm (top) and 550 nm (bottom). (Originally Figure 5 in Zhou et al. (2006). Reproduced with permission)

in the studies highlighted above. In one possible explanation, the "transfer/migration" of the excitation energy to the defect traps must be extremely fast, thus preserving completely the initial polarization of the optical absorption. In such a framework, the somewhat lower r values at shorter excitation wavelengths (Table 3.2) might be understood as being due to some relatively minor depolarization events at the higher-energy trapping sites. The puzzling findings and related explanations are obviously relevant to the mechanistic elucidation of the similarly defect-derived fluorescence emissions in CDots.

The work reported in 2003 by Guldi et al. (2003) was aimed at a more systematic examination of the visible luminescence emissions in dispersed and functionalized SWNTs, with the use of nanotube samples from different production methods including laser ablation (laser-SWNT), arc discharge (arc-SWNT), and HIPco (HIPco-SWNT). While their results were overall consistent with the basic observation and conclusion in the original report by Sun and coworkers (Riggs et al. 2000), there were some additional insights and mechanistic implications discussed in their

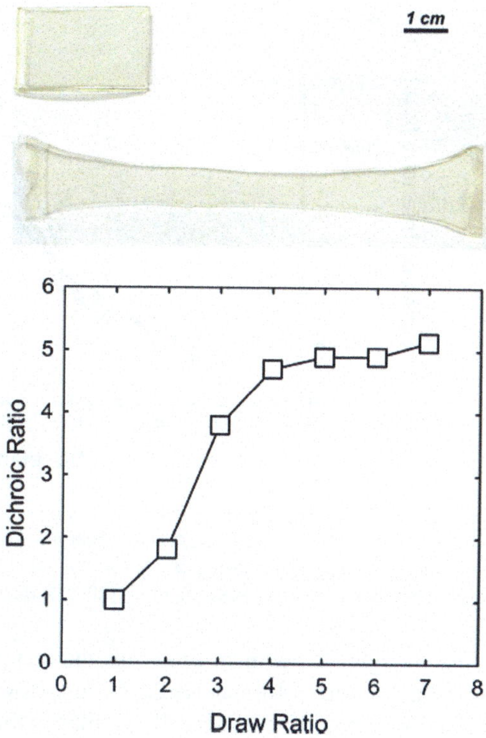

Fig. 3.6 (Upper) PVA films embedded with PPEI-EI-functionalized SWNTs: as-prepared (top) and stretched to a draw ratio of 5 (bottom). (Lower) The observed dichroic ratio for the functionalized SWNTs in stretched PVA film as a function of the draw ratio (500 nm excitation). (Reproduced from Zhou et al. (2006) with permission)

publication (Guldi et al. 2003). For example, as reported, "Upon excitation with visible light, all dispersions/solutions of SWNTs reveal distinct sets of emission spectra. Upon variation of the excitation light, a wavelength progression of the emissive features was seen in most cases. In particular, a red-shift of the excitation wavelength, for instance, from 300 nm to 500 nm led to parallel red-shifts of the emission features. This effect can be as large as 100 nm and depends strongly on the SWNT material. In general, this phenomenon was observed regardless of the class of SWNT—arc-SWNT, HIPco-SWNT and laser-SWNT and regardless of the medium." "In terms of emission shifts, the most striking picture emerges from the heavily oxidized arc-SWNTs. Broad emission bands characteristically shift upon varying the excitation wavelength—reminiscent of earlier observations by Sun and coworkers" (referring to (Riggs et al. 2000)). "To substantiate this observation, the excitation spectrum of each emission maximum was recorded in a separate set-up. This led to good agreement between excitation maximum and initial excitation wavelength (not shown)." The authors found it interesting that "Among all samples examined, the pristine laser-SWNT shows the narrowest distribution. In fact, variation of the excitation frequency between 320 nm and 460 nm led to a redshift of only 4 nm (!)" (Fig. 3.7). The results were used in the discussion of two possible mechanistic frameworks for the visible luminescence emissions, "(*i*) the change of diameter of the SWNT/the size of the bundles and/or (*ii*) the structure/defects on the SWNT's surface." The observed small shifts for pristine laser-SWNT (Fig. 3.7)

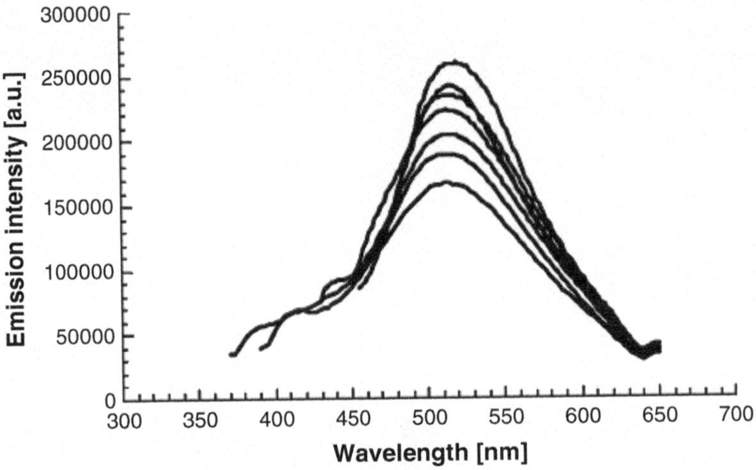

Fig. 3.7 Emission spectra of aqueous suspensions of laser-SWNT with excitation wavelengths of 340, 360, 380, 400, 420, 440, and 460 nm (bottom to top). The emission quantum yield is 0.04. The solutions exhibited absorbances of ~0.4 at the excitation wavelengths; ZnTPP was used as standard. (Originally Figure 1 in Guldi et al. (2003)). Reproduced with permission)

were thought as being in line with the sharp radial-breathing mode peak around 170 cm^{-1} found in Raman spectroscopy. However, looking back, the spectroscopic behavior in Fig. 3.7 was actually similar to that generally found for "naked" carbon nanoparticles, and upon surface functionalization/passivation, spectral shifts and features should become similar to those illustrated in Fig. 3.2. In fact, these authors did observe such changes/effects in triethylene glycol (TEG)-functionalized sample, reporting: "The situation changed dramatically for the TEG-laser-SWNT, which (*i*) gives rise to notable shifts (i.e., 40 nm) and (*ii*) higher quantum yields (i.e., 10 times stronger –0.4%)."

The defect-derived nature of the visible luminescence emissions in functionalized carbon nanotubes (or more broadly carbon nanotubes with various "surface passivation schemes"), as proposed in the early studies discussed above, has remained as a mechanistic framework based mostly on "experimental consistency," namely, the absence of any contradicting experimental evidence. There were also efforts made to probe more directly the functionalized/passivated defect sites in specifically prepared carbon nanotube samples by fluorescence emission measurements at the individual nanotube level. For example, near-field scanning optical microscopy (NSOM) was used to obtain nanoscale images of SWNTs functionalized with derivatized porphyrin molecules (Salazar et al. 2005). The functionalization was such that the carbons at the defect sites of the nanotubes were oxidized into carboxylic acid moieties, which were coupled with the derivatized porphyrin molecules via ester linkages (Fig. 3.8). The porphyrin molecules attached to the defect sites were designed to serve the function of optical tags to enable the probing of the spatial distribution of defect sites on the nanotubes (Fig. 3.8). Both the topography and optical images indicated clustering of the porphyrin tags on specific segments

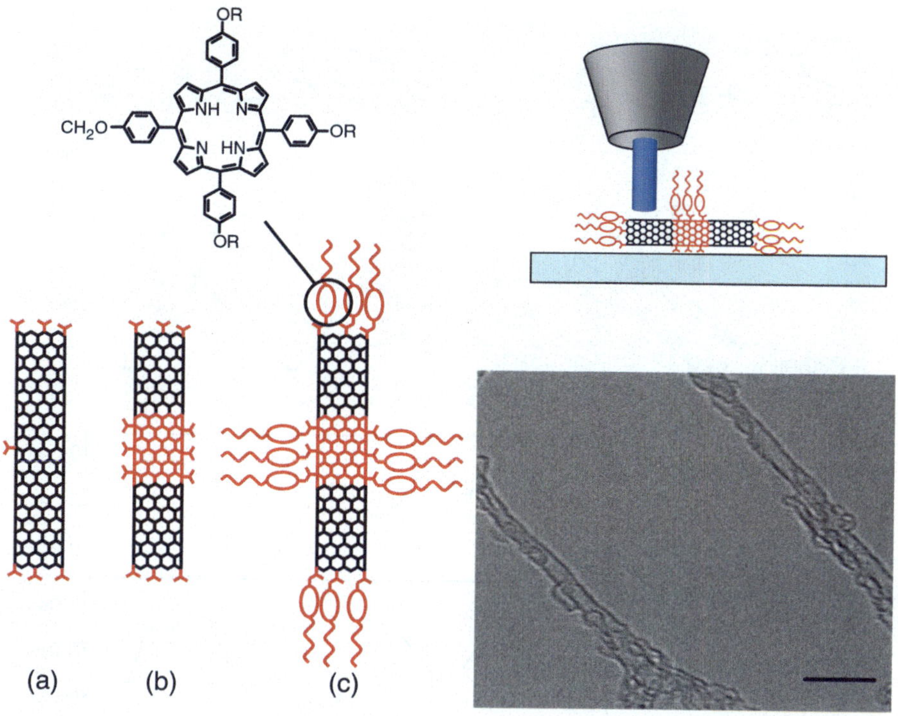

Fig. 3.8 (Left) SWNTs with the carbons at defect sites oxidized and then functionalized/ tethered with porphyrin molecules. (Right-Upper) A cartoon illustration on the NSOM imaging of the porphyrin tagged nanotube. (Right-Lower) A representative TEM image for the functionalized SWNTs (scale bar = 5 nm). (Salazar et al. 2005)

of the nanotube (Fig. 3.9). Images of several functionalized SWNTs ranging from several hundreds of nanometers to one micron in length indicated the presence of short segments (<50 nm) containing an estimated 300–1000 defect site carbons (functionalized, thus tagged by the porphyrin), interspersed with 100–300 nm segments of relatively pristine nanotube (Fig. 3.9). The observed spatial distribution of oxidized defect sites was consistent with "the oxidation proceeding along a reaction front after nucleation at defect sites" (Salazar et al. 2005).

Fast forward to 2010 after CDots already became a popular research topic, the finding made a few years earlier (Sun et al. 2008) that the doping of carbon nanoparticles with metal oxides (such as ZnS and ZnO) could significantly enhance the fluorescence emissions of the resulting CDots prompted a study of using the same doping strategy to probe or "mark" the defect sites on carbon nanotubes (Wang et al. 2010b). The abstract: "Well-dispersed and functionalized carbon nanotubes exhibit visible fluorescence emissions due to passivated defects on the nanotube surface. It was found in this study that the defects in nanotubes could be decorated by an inorganic salt, which augmented the passivation effect of organic functionalization to result in dramatically enhanced emission intensities under both one- and two-photon excitation conditions. The structures and properties of the functionalized

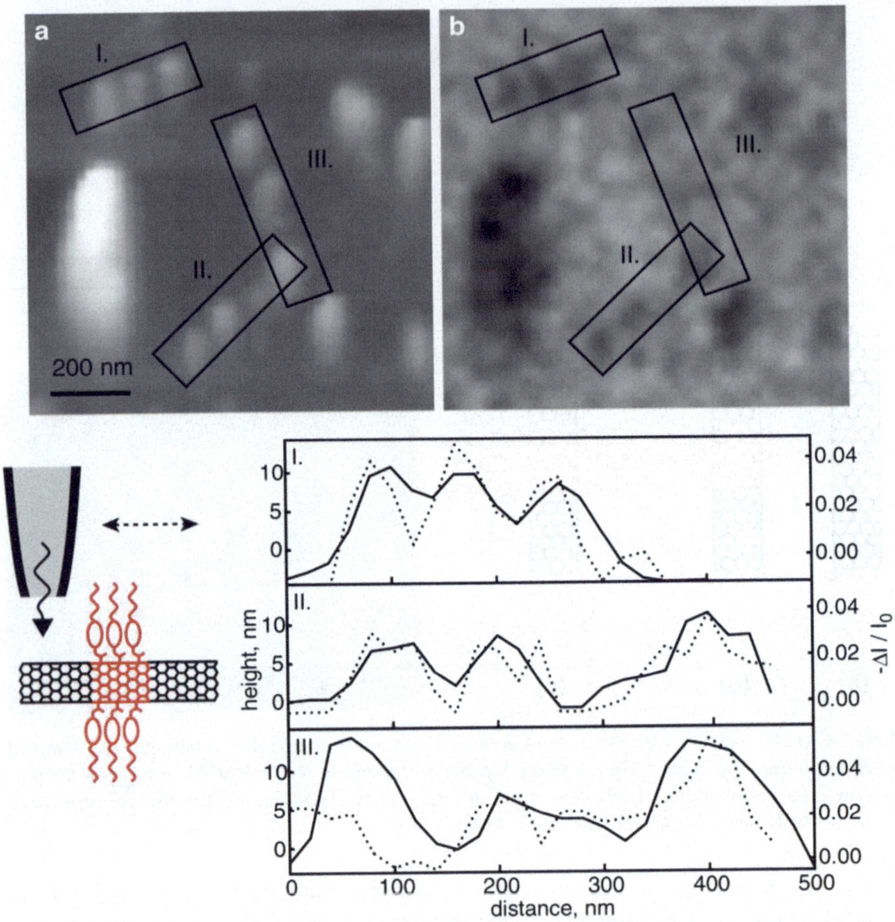

Fig. 3.9 Topography (**a**) and transmission NSOM (**b**) of porphyrin-functionalized/tagged SWNTs. Approximate locations of longer, well-separated tubes are indicated by black boxes. Panels I–III are line scans of topography (solid line) and transmission NSOM (dashed line) for three individual porphyrin-functionalized SWNTs indicated in (**a**). (Salazar et al. 2005)

carbon nanotubes with inorganic coating were thoroughly characterized by using spectroscopy and microscopy techniques. The fluorescence decoration with the coating may serve as a tool in the study of surface defects in carbon nanotubes, and these brightly fluorescent pseudo-one-dimensional nanomaterials may be exploited for optical applications."

The basic idea was such that the ZnS doing of the defect sites (Fig. 3.10) would light up the sites in fluorescence microscopy imaging at submicron resolution (Fig. 3.11), which could be co-localized in atomic force microscopy (AFM, Fig. 3.11) and transmission electron microscopy (TEM, Fig. 3.12) imaging at much higher resolutions.

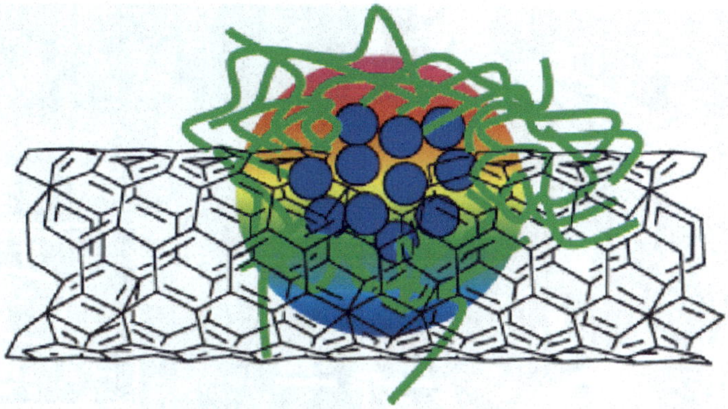

Fig. 3.10 A cartoon illustration on the doping of the carbon nanotube defect site with ZnS and then organic functionalization for enhanced fluorescence emissions. (Reproduced from Wang et al. (2010b) with permission)

On the relevance and purpose of the discussion above concerning the defect-derived photoluminescence emissions in functionalized (loosely defined) carbon nanotubes, these emissions may share mechanistic origins or features with those in CDots. The skepticism and controversies surrounding the originally reported observation of green emissions from the carbon nanotube samples also contributed to or even stimulated the discovery of CDots. In fact, the results reported by Professor Robert Haddon's research group at University of Kentucky and then University of California – Riverside on the chromatography separation of the functionalized carbon nanotube sample into nanotube and "carbonaceous impurity" fractions (Niyogi et al. 2001; Zhao et al. 2001) prompted friendly debates and more experimental investigations that led to the search for and then discovery of CDots by the Sun research group at Clemson University (Sun et al. 2006; Sun 2010).

1.2 "Carbonaceous Impurities"

In the study of green photoluminescence emissions from well-functionalized carbon nanotubes by Sun and coworkers (Riggs et al. 2000; Zhou et al. 2006), the focus was sharply on making sure that the functionalized and solubilized carbon nanotubes were indeed photoluminescent. However, the reviewer comments and rebuttals in the reviewing of that original 2000 paper and the skepticism in the relevant research community immediately following the publication of that paper made the Sun research group at Clemson University realize that similarly functionalized carbon nanoparticles might also be photoluminescent, because the surface defects in the carbon nanoparticles should be no different from those found in the processed and dispersed carbon nanotubes. Thus, they started to look into the real possibility that

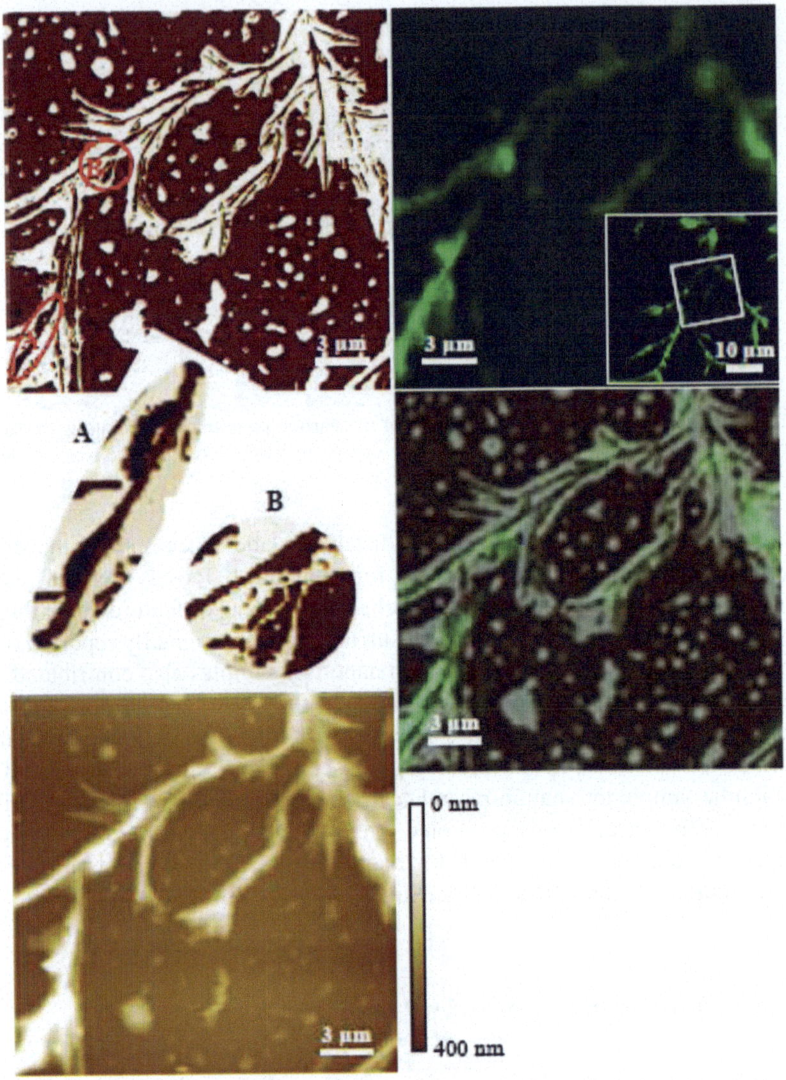

Fig. 3.11 The PEG-MWNT/ZnS specimen on cover glass: the AFM phase (upper left, with the highlighted regions enlarged in middle left) and height (lower left) images, fluorescence image at 458 nm excitation (upper right, enlarged from the highlight region in the inset), and the overlapping between the AFM phase and fluorescence images (lower right). (Originally Figure 3 in Wang et al. (2010b). Reproduced with permission)

nanoscale "carbonaceous impurities" upon the same surface chemical functionalization could be similarly photoluminescent in the visible spectral region. Their effort was further stimulated by the subsequently published studies by Haddon and coworkers (Niyogi et al. 2001; Zhao et al. 2001) and the private communications with Professor Robert Haddon at University of California – Riverside on the subject.

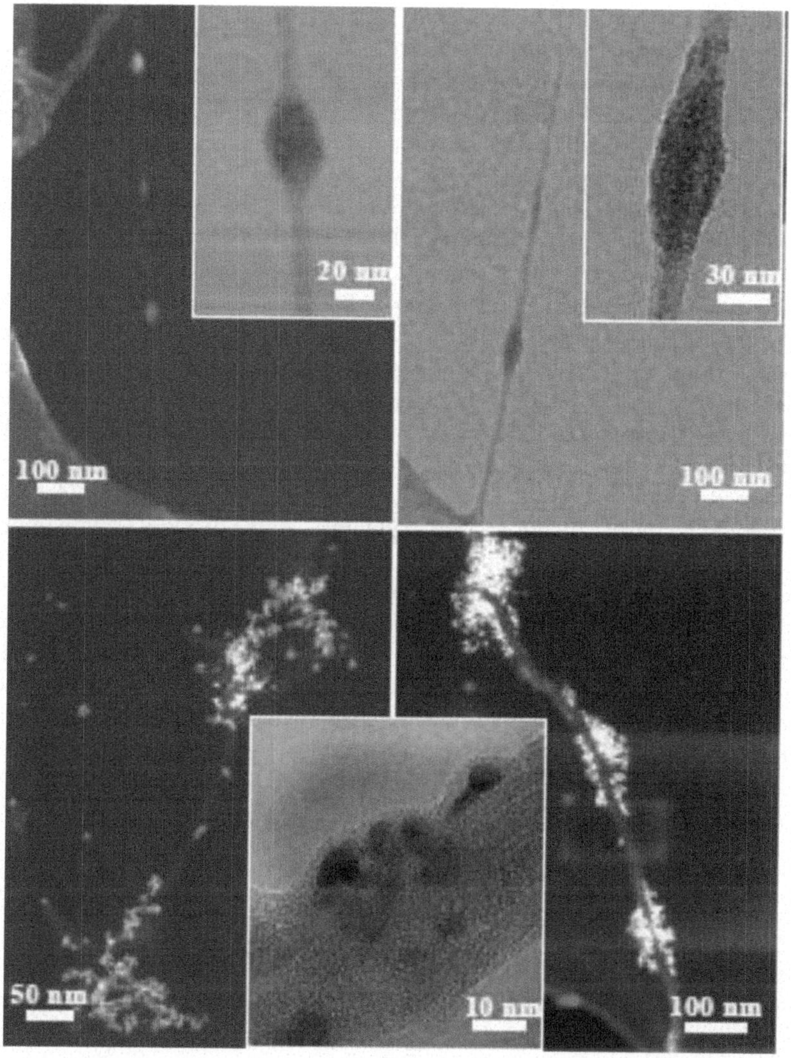

Fig. 3.12 TEM images of the specimens on holey carbon-coated copper grids. Upper: PEG-SWNT/ZnS (left, scanning-TEM Z-contrast dark field; right, TEM bright field; insets, corresponding high-resolution images). Lower: PEG-MWNT/ZnS (left and right, scanning-TEM Z-contrast dark field; inset, TEM bright field at high resolution). (Originally Figure 4 in Wang et al. (2010b). Reproduced with permission)

In the two papers published in 2001 (Niyogi et al. 2001; Zhao et al. 2001), Haddon and coworkers reported some interesting findings as by-products in their studies of octadecylamine-functionalized SWNT samples. While their primary purpose was to fractionate the samples on gel permeation chromatography columns to harvest clean SWNTs, some of their reported results and associated descriptions and conclusions were relevant to the active investigation at the time by the

Sun research group at Clemson University on the colorful photoluminescence emissions from chemically functionalized nanoscale carbonaceous materials in solution. For example, Haddon and coworkers reported that the octadecylamine-functionalized SWNT sample could be separated on special gel permeation chromatography columns to obtain the first fraction containing "octadecylamine-functionalized SWNTs of high purity" and then one or two more other fractions, depending on the gel columns used. These other fractions were found to be photoluminescent in the visible spectrum (Fig. 3.13) (Zhao et al. 2001).

Inspired by the studies of Haddon and coworkers discussed above, the Sun research group at Clemson University adopted the same fractionation approach and added more findings and understandings on top of those reported by Haddon and coworkers. Some of the additional findings and understandings were highlighted in the 2005 *Journal of Physical Chemistry* paper (Lin et al. 2005). One such finding was that the first separated fraction of the octadecylamine-functionalized SWNT sample, containing the functionalized SWNTs of high purity and nonluminescent (thus agreeing with the results reported by Haddon and coworkers (Zhao et al. 2001)), could be further functionalized to obtain a sample of SWNTs with improved surface defect passivation and similarly bright visible photoluminescence emissions (Lin et al. 2005). The original purpose of that investigation was to use the functionalized carbon nanotube sample as a source for functionalized carbon nanoparticles (or generally "carbonaceous impurities"), which could then be studied more systematically and quantitatively for their photoluminescence properties. However, in light of the results from the further functionalization of the first fraction from the gel column separation, the project team became seriously concerned about the possible cross-contamination between the functionalized carbon nanotubes and carbon

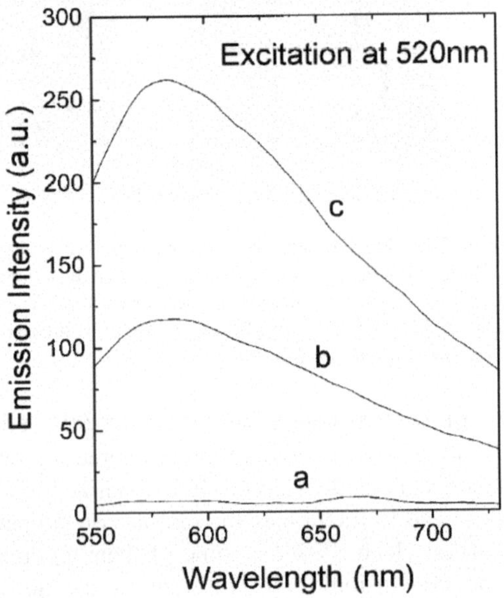

Fig. 3.13 Emission spectra of octadecylamine-functionalized SWNTs (in THF solution, separated by a Styragel HMW7 column), where a, b, and c refer to fraction 1, the starting material, and fraction 2, respectively. (Originally Figure 6 in Zhao et al. (2001). Reproduced with permission)

nanoparticles in the various separated fractions, concluding that the results from the approach of fractionating functionalized carbon nanotube samples would not be as "absolutely unambiguous" as desired to address the skepticisms and confusions in the relevant research community at the time. It was decided that the objective for the project to produce absolutely clear evidence for an indisputable conclusion could only be achieved by using specifically prepared "pure carbon nanoparticles" without any chance for contamination by carbon nanotubes or the like. Nevertheless, one of the project team members who was still a freshman graduate student subsequently moved to another institution and became the primary author for a 2004 *Journal of American Chemical Society* publication on the same subject.

At Clemson University, the Sun research group did pursue the objective for absolutely unambiguous evidence by setting up a laser ablation apparatus (Fig. 2.14) to produce carbon nanopowder samples for pure carbon nanoparticles, which led to the indisputable conclusion that the chemical functionalization of small carbon nanoparticles could yield "dots" of bright and colorful photoluminescence emissions comparable to those exhibited by conventional semiconductor QDs, thus the invention of carbon dots (CDots) (Sun et al. 2006; Sun 2010).

2 The Discovery

Somewhat surprising in the search and selection of a production method for small nanoparticles of neat carbon was the lack of choices, let alone simple and inexpensive. In the end, the method reported in the literature on using laser ablation of a highly pure graphite target in the presence of water vapor was adopted for the preparation of carbon soot with a relatively higher content of small carbon nanoparticles (Fig. 2.14) (Sun et al. 2006).

The as-produced carbon soot sample was processed in a protocol similar to what was developed for the purification and dispersion of carbon nanotubes to harvest surface-oxidized small carbon nanoparticles in a stable aqueous suspension. The assumption behind the processing was such that there could be roughly three different outcomes for likely a mixture of various carbon nanoparticles in the carbon soot sample: (1) the really small ones being destroyed (oxidized into carbon dioxide); (2) the larger ones being precipitated in the centrifugation; and (3) the small nanoparticles but not too small being survivors with surface oxygen-containing moieties like carboxylic acids. Indeed, the aqueous dispersed carbon nanoparticles thus harvested were found to be small, on the order of 5 nm in diameter. Results from the characterization of these small carbon nanoparticles, especially the NMR measurements of those from the processing of ^{13}C-enriched carbon soot sample, confirmed the presence of surface carboxylic acid moieties.

The chemical functionalization of the small carbon nanoparticles bearing surface carboxylic acid groups was the use of amino molecules and polymers under amidation reaction conditions (Fig. 3.14), where the thinking behind the use of highly soluble amino polymers was to provide enough "solubilization power" to ensure

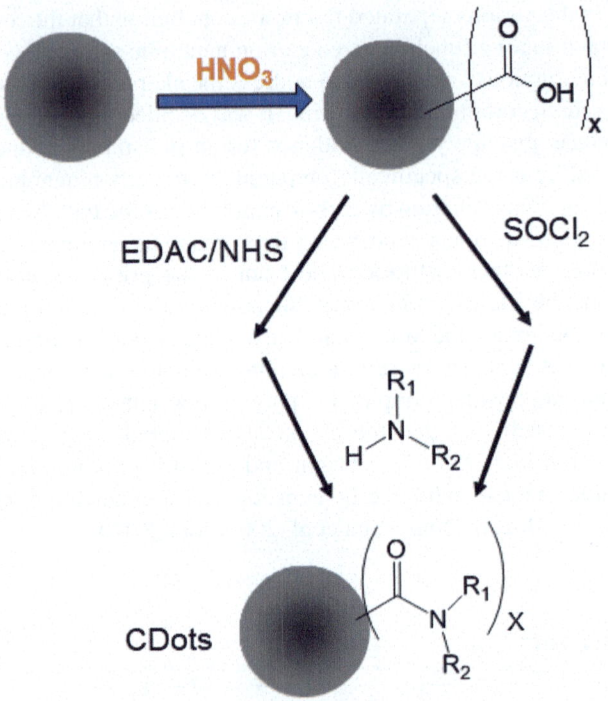

Fig. 3.14 The amidation of carbon nanoparticle surface-bound carboxylic acids with amine molecules in two different reaction schemes. Right: the treatment with thionyl chloride ($SOCl_2$) to form acyl chloride for ready substitution by the amine molecule. This scheme is associated with relatively mild reaction conditions, hardly capable of producing any dyes or causing meaningful carbonization. Left: the coupling of carboxylic acids and amines with the mediation/activation by the commonly used carbodiimide (EDAC). This scheme is designed to be very "gentle," commonly used with fragile biological species

that the resulting CDots would be readily soluble to form "true solutions" (Sun et al. 2006). Years later, however, it became clear based on significant experimental results acquired since 2006 that the solubilization concern in the original finding and the subsequent initial development of CDots was not necessary, as the "solution quality" for CDots of surface functionalization by some small molecules could still be excellent (LeCroy et al. 2014; Yang et al. 2016). The initially targeted surface functionalization structure was the formation of amide linkages, though according to subsequent investigations the functionalization of the carbon nanoparticle surface might be more like a soft shell or corona filled with organic species used for the functionalization (Fig. 1.1).

The observed absorption spectrum of the CDots (Fig. 1.2) was rather similar to that of the dispersed small carbon nanoparticles before the surface functionalization, namely, no significant changes to the electronic transitions of the core carbon nanoparticles (due to π-plasmons) after the surface functionalization with molecules optically transparent in the near-UV and visible spectral regions. However, more interesting and important were the observed bright and colorful photoluminescence emissions at different excitation wavelengths (Fig. 1.2), significantly different from those of the aqueous dispersed small carbon nanoparticles.

The strong photoluminescence (or fluorescence as casually referred to in many publications without any strict consideration of the electronic states involved) emissions were comparable with those found in functionalized carbon nanotubes, which as discussed above actually prompted the search for the same or similar emission properties in functionalized carbon nanoparticles. The characteristic excitation wavelength dependence of the emissions demonstrated in Fig. 1.2 was phenomenologically similar to that known for conventional semiconductor QDs, which was behind the designation of the functionalized small carbon nanoparticles as "dots," despite the major differences both spectroscopically and mechanistically. Among the spectroscopic differences, unlike in the semiconductor QDs with different absorption spectra and fluorescence colors corresponding to dots of different sizes, the same CDots sample of a single absorption spectrum could produce different emission colors with excitations into different wavelengths of the absorption spectrum. The mechanistic explanation in the original report on the observed photoluminescence emission properties of CDots was their being "defects-derived" (Sun et al. 2006), different from the bandgap nature of fluorescence emissions in conventional semiconductor QDs. In fact, the mechanism focusing on functionalized defects was fundamentally similar to what was proposed a few years earlier in the rationalization of similar photoluminescence properties found in functionalized and solubilized carbon nanotubes (Riggs et al. 2000). Since 2006, the same mechanistic framework has been adopted in the CDots research field, despite the seemingly pretty desperate effort in some corner of the field on linking CDots to graphenes by essentially "hijacking" the original concept and definition proposed by several research groups (e.g., (Ponomarenko et al. 2008; Westervelt 2008)) on "graphene quantum dots" (see more detailed discussion in Chap. 5).

It was found at almost the same time as the discovery that CDots could be excited by two near-IR photons from a femtosecond pulsed laser for similarly bright and colorful photoluminescence emissions in the visible, namely, CDots of excellent multiphoton photoluminescence properties (Cao et al. 2007). Experimentally, the CDots functionalized with poly(propionylethylenimine-*co*-ethylenimine) (PPEI-EI, with EI fraction ~ 20%) were measured by using a Leica confocal fluorescence microscope equipped with an argon ion laser and a femtosecond pulsed Ti:Sapphire laser. The CDots were found to be strongly emissive in the visible spectrum with either the argon ion laser excitation (458 nm) or the femtosecond pulsed laser for two-photon excitation in the near-IR (800 nm). As compared in Fig. 3.15, the one- and two-photon photoluminescence images for the same scanning area matched well. The CDots were photostable under the two-photon imaging conditions (upon repeated 800 nm excitations equivalent to generating the image in Fig. 3.15 for at least 3000 times, no meaningful changes in emission intensities). The two-photon emission spectra measured on the same microscope setup were found to be similar to the one-photon spectra of the same CDots (Fig. 3.16).

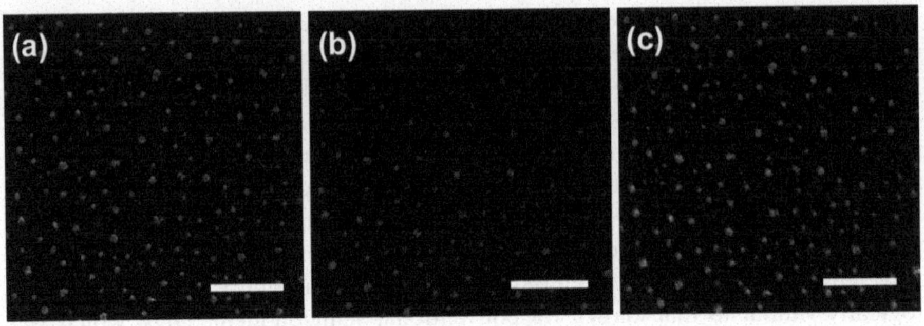

Fig. 3.15 Luminescence images (all scale bars 20 μm) of the CDots with (**a**) argon ion laser excitation at 458 nm and (**b**) femtosecond pulsed laser excitation at 800 nm; (**c**) is an overlay of (**a**) and (**b**). (Originally Figure 2 in Cao et al. (2007). Reproduced with permission)

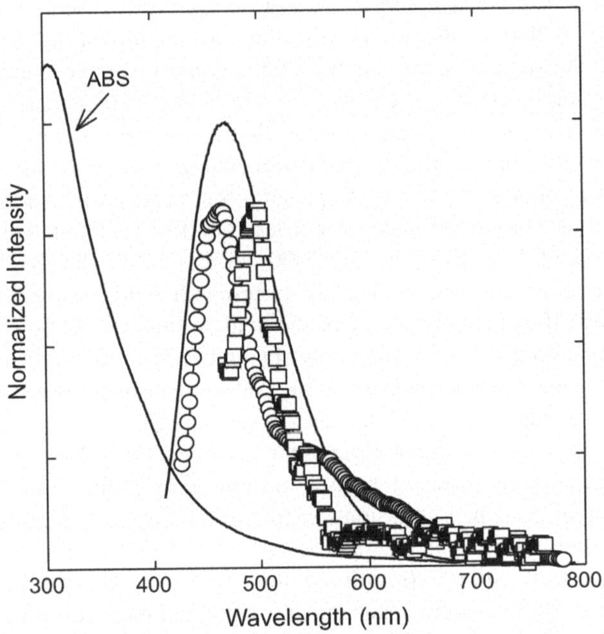

Fig. 3.16 The one-photon (-□-, 458 nm excitation) and two-photon (-O-, 800 nm excitation) luminescence spectra of the CDots on glass substrate (prepared with infinite dilution, and optical spot diameter ~500 nm, covering multiple dots immobilized on the substrate) are compared with solution-phase absorption (ABS) and luminescence (solid line, 400 nm excitation) spectra. (Originally part of Figure 3 in Cao et al. (2007). Reproduced with permission)

The two-photon in nature for the photoluminescence with excitation in near-IR by the pulsed laser was confirmed by the dependence of observed emission intensities on the excitation laser power, namely, the obvious quadratic relationship between the excitation laser power and the photoluminescence intensity (Fig. 3.17).

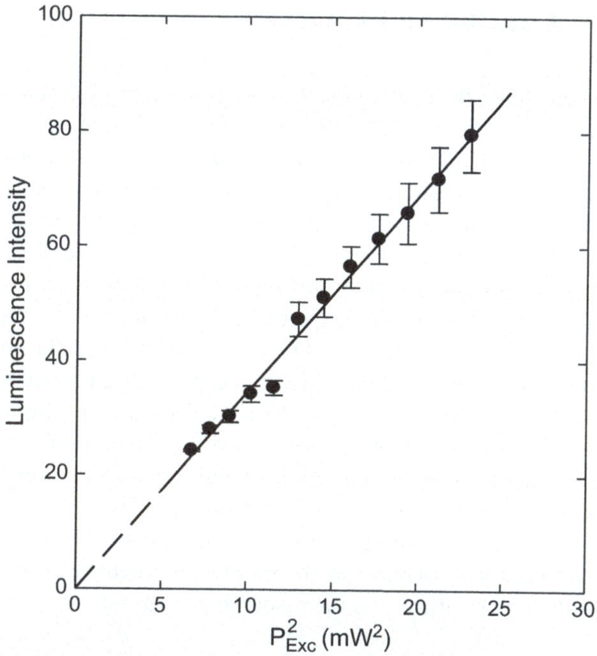

Fig. 3.17 The quadratic relationship of the observed two-photon luminescent intensity of the CDots on glass substrate with the excitation laser power at 800 nm (P_{Exc}, as measured at the focal plane). (Originally part of Figure 3 in Cao et al. (2007). Reproduced with permission)

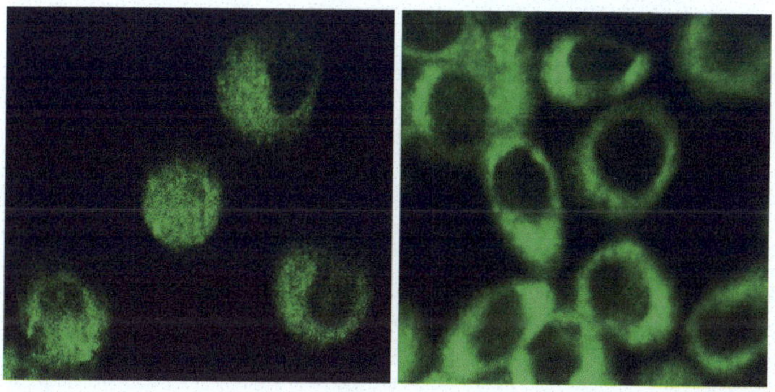

Fig. 3.18 Representative two-photon luminescence images (two-photon 800 nm excitation) of human breast cancer MCF-7 cells with internalized CDots. (Originally Figure 4 in Cao et al. (2007). Reproduced with permission)

The estimated two-photon absorption cross-section of CDots at 800 nm was 39,000 ± 5000 GM (Goeppert-Mayer unit, with 1 GM = 10^{-50} cm^4 s/photon), making CDots highly competitive to other two-photon fluorescence nanomaterials, such as CdSe and CdSe/ZnS QDs. Since the original report (Cao et al. 2007), the high performance of CDots as multiphoton fluorescence probes has generally been validated ((Kuo et al. 2016), for example). A major application of high-performance multiphoton probes is in bioimaging, as highlighted in the original report (Fig. 3.18) (Cao et al. 2007).

3 Concluding Remarks

There have been misunderstandings in some literature reports that "carbon dots were found by accident." In reality, no accident at all, as made evident by the historical account presented in this chapter. To be fair, however, the finding of defect-derived photoluminescence emissions in functionalized carbon nanotubes was indeed accidental. In fact, for that finding reported in (Riggs et al. 2000), the project team was initially "terrified," because there were absolutely no reasons to expect any photoluminescence emissions of carbon nanotubes in the visible spectrum (fully aware of the bandgaps being in the near-IR beyond the coverage of the available spectrometer to only 1100 nm). Even when the manuscript was accepted for publication by the *JACS* editor, the project team wanted to hold it for as long as possible/allowed and to use the time for another careful check on all possible artifacts. It was then that the project team came to realize the likelihood of functionalized small carbon nanoparticles being photoluminescent in the visible spectrum as well, because the chemical functionalization in both nanotubes and nanoparticles was associated with the same kind of defects. After that, it was simply to do whatever necessary to build an absolutely convincing case on carbon dots to be presented to the patent office and the research community.

Chapter 4
Carbon Dots: The Classical Synthesis and Properties

The structure of CDots is built upon the template of small carbon nanoparticles, each of which is surface "passivated" in the synthesis to become the core in a resulting dot. The carbon nanoparticles used for CDots have so far been generally less than 10 nm in diameter and largely amorphous, corresponding to broad or structure-less X-ray powder diffraction patterns. However, being amorphous is hardly a requirement or even an advantage in terms of the properties and performance of the corresponding CDots. In fact, more graphitic and crystalline carbon nanoparticles or others as described in Chap. 2 can also be surface passivated by chemical functionalization or the like to form CDots or equivalent nanostructures for the same optical and redox characteristics (Sun 2010; Liang and Sun 2019). With such a dot structure, logically CDots are generally defined as "small carbon nanoparticles with various surface passivation schemes" (Fig. 1.1).

1 The Classical Synthesis

For the synthesis of CDots, the use of pre-existing carbon nanoparticles for surface passivation, mostly through the functionalization with organic molecules or other species, obviously represents a more reliable and "safer" approach, with the resulting dots structurally more closely aligned with the general definition. Correspondingly, the observed optical and other related properties of these dot samples should be considered as being more genuine to the CDots as defined. The approach based on the surface passivation of pre-processed and selected small carbon nanoparticles has been designated as the deliberate chemical functionalization method. It may be considered as the "classical" synthesis because of its more literally adhering to the definition of CDots.

© Springer Nature Switzerland AG 2020
Y.-P. Sun, *Carbon Dots*, https://doi.org/10.1007/978-3-030-41184-8_4

1.1 *Small Carbon Nanoparticles*

The chemical functionalization synthesis starts with the processing of carbon soot samples from various production methods or commercially acquired carbon nanopowder samples, which typically involves the oxidative acid treatment under refluxing conditions. Among purposes of the treatment are the breaking up of nanoparticle aggregates and the creation of oxygen-containing moieties such as carboxylic acids on the nanoparticle surface, thus facilitating their solvent dispersion. An implicit assumption, which has largely been validated experimentally, is that smaller carbon nanoparticles populated with the oxygen-containing moieties on the surface are more solvent dispersible to allow their being harvested in a relatively simple protocol. For the originally reported CDots (Sun et al. 2006), the small carbon nanoparticles were harvested from carbon soot samples produced in laser ablation of a carbon target in water vapor, which was designed to have the resulting soot samples contain a larger population of smaller carbon nanoparticles. Since then, many other carbon soot materials containing nanoparticles have been used as precursors in the protocol for the harvesting of surface-oxidized small carbon nanoparticles, as discussed in Chap. 2. These carbon soot materials vary in morphology according to X-ray powder diffraction results, including carbon soot samples from arc discharge of graphite rods (largely amorphous to partially graphitic), some selected carbon nanopowder samples acquired from an established chemical supplier (sourced from laser ablation according to the supplier, partially graphitic), soot samples from the carbonization of submicron polypyrrole particles in argon atmosphere at 800 °C (slightly graphitic), and other commercially supplied carbon nanopowders from carbonization of plant materials (largely amorphous). The small carbon nanoparticles from all these sources have been functionalized by selected organic molecules in various experimental protocols under different conditions for CDots of varying fluorescence emission properties, but there has been no evidence suggesting any significant and consistent dependence on the sources of the small carbon nanoparticles. This is particularly meaningful with the use of carbon nanoparticles sourced from the carbonization of submicron polypyrrole particles, in which the nitrogen content is obviously substantial. Thus, the results, among others, suggest strongly the insensitivity of CDots to the nitrogen doping in the core carbon nanoparticles, which may have major mechanistic implications.

1.2 *Functionalization via Amidation and Dot Characterizations*

The functionalization chemistry for the pre-processed and selected small carbon nanoparticles has mostly been amidation, targeting the carboxylic acid groups on the nanoparticle surface (Figs. 3.14, 4.1). Aminopolymers such as PPEI-EI and oligomers of relatively large molecular weights (such as PEG_{1500N}, Fig. 4.1) were

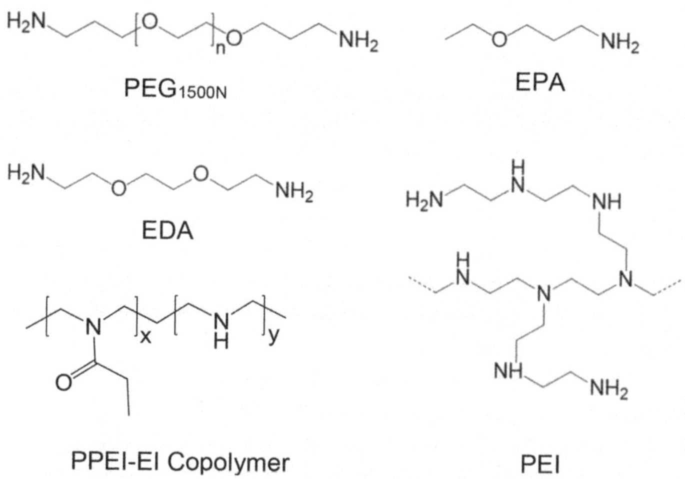

Fig. 4.1 Molecules and polymers bearing amine groups used for the functionalization of pre-processed and selected small carbon nanoparticles for CDots. PEG_{1500N} = oligomeric polyethylene glycol diamine of average molecular weight ~ 1500; EPA = 3-ethoxypropylamine; EDA = 2,2′-(ethylenedioxy)bis(ethylamine); PEI = oligomeric polyethylenimine (branched) of average molecular weight ~ 600 or 1200; PPEI-EI = poly(propionylethylenimine-*co*-ethylenimine)

initially selected due to the consideration that the functional molecules should provide sufficient solubility to "drag" their functionalized carbon nanoparticles into the solvent system for the formation of close to true solutions of the CDots (Sun et al. 2006). Indeed, the corresponding CDots were found to be readily soluble in water and some other solvents for solutions that exhibited no significant light scattering effect, as the substantial scattering effect often serves as an indicator for poor solution quality in the solubilization of nanomaterials. However, rather surprising and also very interesting and valuable was the finding that small amine molecules such as 2,2′-(ethylenedioxy)bis(ethylamine) (EDA, molecular weight 148) (LeCroy et al. 2014) and 3-ethoxypropylamine (EPA, molecular weight 103) (Yang et al. 2016) (Fig. 4.1) could also be used in the amidation reaction to functionalize effectively the pre-processed and selected small carbon nanoparticles for CDots of similarly excellent solubility and also high fluorescence emission performance. Their solution characteristics and observed optical absorption and fluorescence emission properties were found to be comparable to those of the CDots with functionalization by polymers or larger oligomers. The successful preparation of EDA-CDots and EPA-CDots and their obviously competitive performance have demonstrated not only the versatility with respect to compositions in the surface passivation layer of CDots but also opportunities for employing a wide variety of molecules and species for the effective surface passivation via chemical functionalization.

The chemical functionalization reaction conditions for amidation (referring specifically to those involving the treatment with thionyl chloride, Fig. 3.14) are generally rather mild, with the applied temperature (up to 120 °C) insufficient for any meaningful carbonization products. The amidation conditions are also known as being specific, without any possibility for side reactions that would lead to the formation of dye species capable of the observed substantial absorption and fluorescence emission behaviors in the visible spectral region, extending into the near-IR. These considerations and conclusions were related to the initial concerns on the observed properties of CDots being due to some artifacts, which have since been put to rest. Nevertheless, the considerations discussed above are still relevant to some current issues in the CDots research field, including those associated with more recent debates on the composition and quality of dot samples from the "one-pot" carbonization synthesis method, as discussed in more detail in Chap. 5.

Historically on addressing the initial concerns in the research community on potential artifacts (soon after the original report on CDots), Dr. Lingrong (Linna) Gu and Dr. Liangwei Qu, who were both graduate students at the time in the Sun research group at Clemson University, performed a careful investigation on a different amidation approach that is associated with considerably milder reaction conditions. In fact, their selected amidation approach based on the use of a carbodiimide activation agent, specifically 1-ethyl-3-(3-dimethylaminopropyl)carbodiimide (EDAC), is widely employed in the literature for the coupling of fragile biological species such as antibodies. They adopted the experimental protocol originally developed for the attachment of proteins to carbon nanotubes (Huang et al. 2002). The CDots from such an amidation synthesis exhibited largely the same absorption and fluorescence emission spectra, but the observed fluorescence quantum yields were lower than those of the dot samples prepared by the amidation with thionyl chloride (Fig. 3.14), suggesting less effective surface passivation. Nevertheless, the results did serve to rule out any possibilities for the observed fluorescence properties being associated with non-CDots artifacts.

Similarly, the mild reaction conditions in amidation are unlikely to alter the structural and other characteristics of the small carbon nanoparticles post functionalization, such as the particle sizes, compositions, and morphology. On the sizes of the nanoparticles before and after the functionalization, for example, available experimental results based on microscopy imaging techniques have suggested no substantial changes, though more quantitative analyses at the individual dot level still have room for improvements. For size analyses of CDots, transmission electron microscopy (TEM) and atomic force microscopy (AFM) have commonly been used, such as those for EDA-CDots shown in Fig. 4.2.

Since the overwhelming majority of the reported CDots from the deliberate chemical functionalization synthesis have been based on the use of amine molecules for the functionalization under amidation reaction conditions, various characterization methods including especially ^{1}H and ^{13}C NMR measurements and FT-IR have been employed, yielding results that are generally consistent with the targeted functionalization. For example, shown in Fig. 4.3 are NMR results for EDA-CDots (LeCroy et al. 2014). FT-IR results have not been as informative in the probing and understanding of the functionalization, but they have so far served the purpose of eliminating any significant presence of other species in the CDots samples from the functionalization reactions.

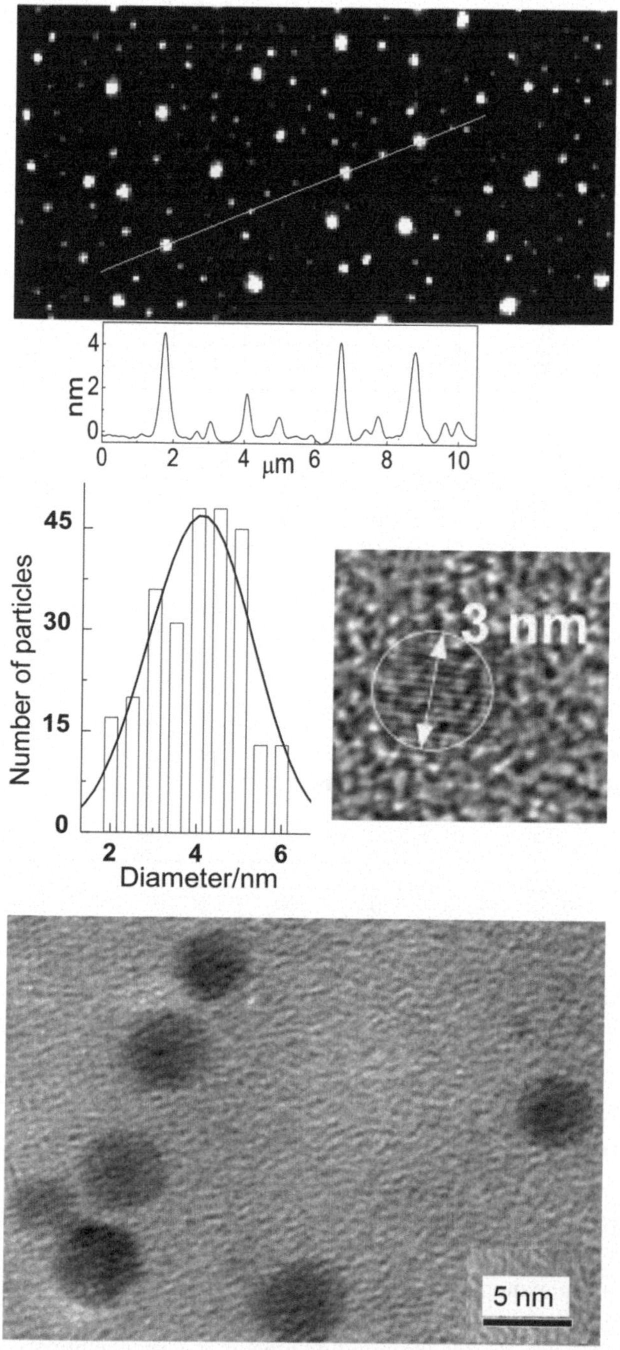

Fig. 4.2 Top, atomic force microscopy (AFM) imaging results for EDA-CDots on mica substrate, with height profiles of some dots along the line highlighted; middle, (left) the size distribution based on height analyses of multiple AFM images, fitted with the Gaussian distribution curve and (right) a high-resolution transmission electron microscopy (TEM) image illustrating the carbon core in a dot; bottom, TEM images of the EDA-CDots. (Originally Figure 3 in LeCroy et al. (2014). Reproduced with permission)

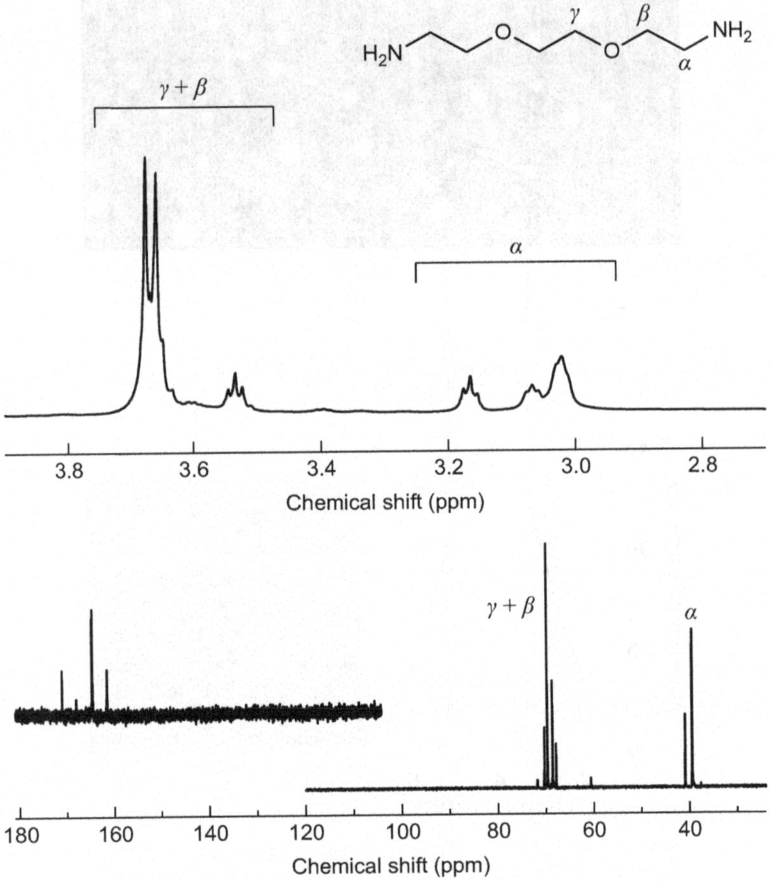

Fig. 4.3 Upper: ^{1}H-NMR spectrum of EDA-CDots. Lower: ^{13}C-NMR spectrum of EDA-CDots and (inset on the left) the lower-field portion of the spectrum for the ^{13}C-enriched EDA-CDots. (Originally Figure 4 in LeCroy et al. (2014). Reproduced with permission)

2 Absorption and Fluorescence Properties

As discussed above, the mild reaction conditions in the chemical functionalization synthesis of CDots should not affect the structural characteristics and properties of the small carbon nanoparticles, including apparently their optical absorptions. Shown in Fig. 4.4 is a general agreement between observed absorption spectra of CDots in solution and their corresponding precursor carbon nanoparticles in a stable dispersion. There are no quantitative data available on possible changes in absorptivities from the carbon nanoparticles to CDots, mostly due to technical difficulties in the measurements for such quantitative data, though the empirical results accumulated so far seem to suggest that if there are changes, they are unlikely to be

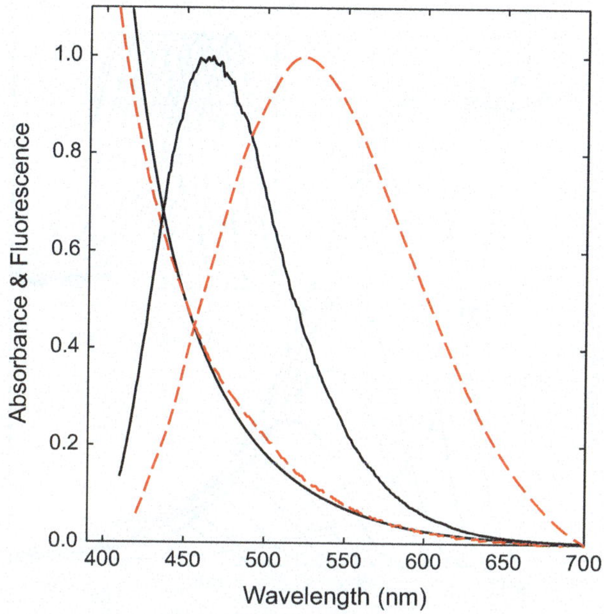

Fig. 4.4 Absorption (ABS) and normalized fluorescence (FLSC, 400 nm excitation) spectra of EDA-CDots in aqueous solution (solid line) and aqueous dispersed carbon nanoparticles (dashed line). (Originally Figure 3 in LeCroy et al. (2017b))

dramatic. The results also suggest that the odds are probably in favor of some slight increases in the optical absorptivities from the precursor carbon nanoparticles to CDots, but no more than 50% for sure based again on available empirical data.

CDots are characterized by their bright and colorful fluorescence emissions (Fig. 1.2). In a direct comparison between fluorescence spectra of the small carbon nanoparticles and their corresponding CDots at the same wavelength for excitation, the latter was found as being blue-shifted with a change in spectral shape (Fig. 4.4), accompanied by more than an order of magnitude increases in the emission intensities (LeCroy et al. 2017b). The results could be rationalized as such that the inherent nature of the emissive states is not the same in the two systems (LeCroy et al. 2017b).

Both fluorescence spectra and quantum yields of CDots are excitation wavelength dependent in a rather characteristic pattern. In fact, such a pattern has been observed across CDots with many different surface functionalization molecules, despite the fact that the absolute values of fluorescence quantum yields might be different between the different CDots. For EDA-CDots as an example (LeCroy et al. 2017b), the excitation wavelength dependence of the observed fluorescence spectra is essentially the same as that originally found for PPEI-EI-CDots (Fig. 1.2) (Sun et al. 2006). As shown in Fig. 4.5, the fluorescence band maximum shifts progressively with the excitation toward longer wavelengths, accompanied by progressive narrowing of the emission bandwidth and decreasing of emission intensities. These dramatic changes of the emission color as a function of the excitation wavelength

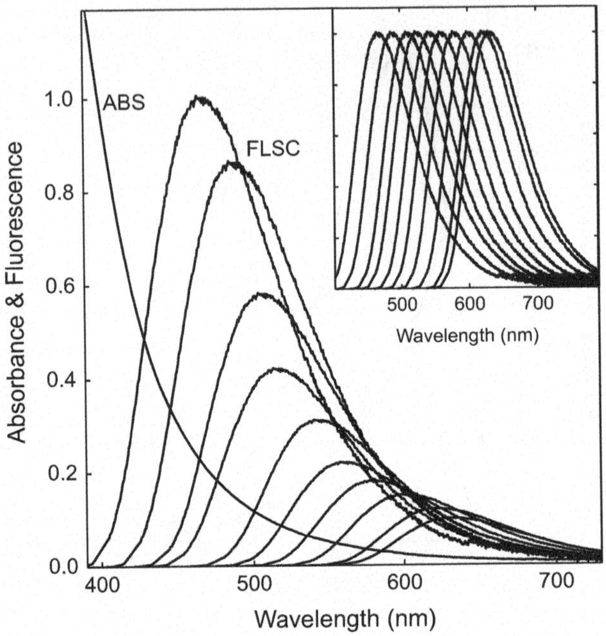

Fig. 4.5 Absorption spectrum (ABS) and fluorescence spectra (FLSC) of EDA-CDots in aqueous solution excited at (in the order of progressively lower peak intensity) 400–580 nm in 20 nm increment. The corresponding normalized FLSC spectra are shown in the inset. (Originally Figure 4 in LeCroy et al. (2017b). Reproduced with permission)

likely reflect the availability of surface traps at different energies, probably variable from dot to dot as a consequence of structural inhomogeneity. Interestingly the characteristic dependence of fluorescence quantum yields on excitation wavelengths, namely, higher quantum yields in the blue and progressively lower at longer excitation wavelengths, actually tracks closely the absorption spectral profile (Fig. 4.6). Even more interesting and mechanistically significant is that the excitation wavelength dependence of fluorescence quantum yields (Φ_F) in the CDots is decoupled from that of observed fluorescence decays and lifetimes (Fig. 4.6, Table 4.1). Such behaviors, apparently shared by other CDots, have major mechanistic implications.

2.1 Mechanistic Framework

In the original report on CDots (Sun et al. 2006), the photoluminescence emissions were considered mechanistically as being "defect-derived" in nature. The proposed mechanistic framework has since been adopted in the research field, for which a cartoon illustration is shown in Fig. 4.7 with the description as follows. The photoexcitation of core carbon nanoparticles in CDots results in efficient charge

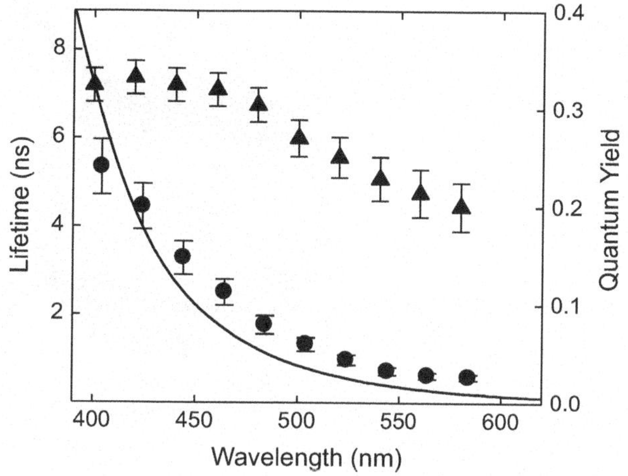

Fig. 4.6 The absorption spectrum (solid line), fluorescence quantum yields (Φ_F, solid circle), and averaged fluorescence lifetimes (τ_F, solid triangle) of EDA-CDots in aqueous solution at different excitation wavelengths. (Originally part of Figure 5 in LeCroy et al. (2017b). Reproduced with permission)

Table 4.1 Fluorescence decay kinetics of EDA-CDots deconvoluted with a biexponential decay function

λ_{EX} (nm)	λ_{EM} (nm)	τ_{F1} (ns)	A_1 (%)	τ_{F2} (ns)	A_2 (%)	$<\tau_F>$ (ns)
400	465	2.2	25	7.7	75	7.2
420	485	2.2	26	7.9	74	7.4
440	510	2.1	29	7.8	71	7.2
460	525	2.0	30	7.7	70	7.1
480	545	1.8	30	7.3	70	6.8
500	565	1.6	31	6.5	69	6.0
520	585	1.5	34	6.1	66	5.6
540	600	1.4	40	5.7	60	5.1
560	620	1.3	43	5.4	57	4.8
580	630	1.1	42	5.0	58	4.5

Originally Table 1 in LeCroy et al. (2017b)
λ_{EX} and λ_{EM} are the excitation wavelength and the monitored emission wavelength, respectively
The average fluorescence lifetime $<\tau_F> = (A_1\tau_{F1}^2 + A_2\tau_{F2}^2)/(A_1\tau_{F1} + A_2\tau_{F2})$

separation (on the time scale of picosecond/sub-picosecond according to available ultrafast transient spectroscopy results), with the electrons and holes (or radical anions and cations in a different description) trapped at various surface sites of the nanoparticles. The surface passivation of the core carbon nanoparticles must be for stabilizing the trapping sites for the electrons and holes, and radiative recombinations of the electrons and holes must be responsible for the observed fluorescence emissions. This is in many respects similar to the photoexcited state mechanism in conventional semiconductor QDs. While the classical quantum confinement found in semiconductor QDs is not applicable in CDots, as the emission colors in the latter are associated with different trapping sites and/or recombination paths of the electrons and holes, generally small sizes for the core carbon nanoparticles are required for an extremely large surface-to-volume ratio to ensure a large population

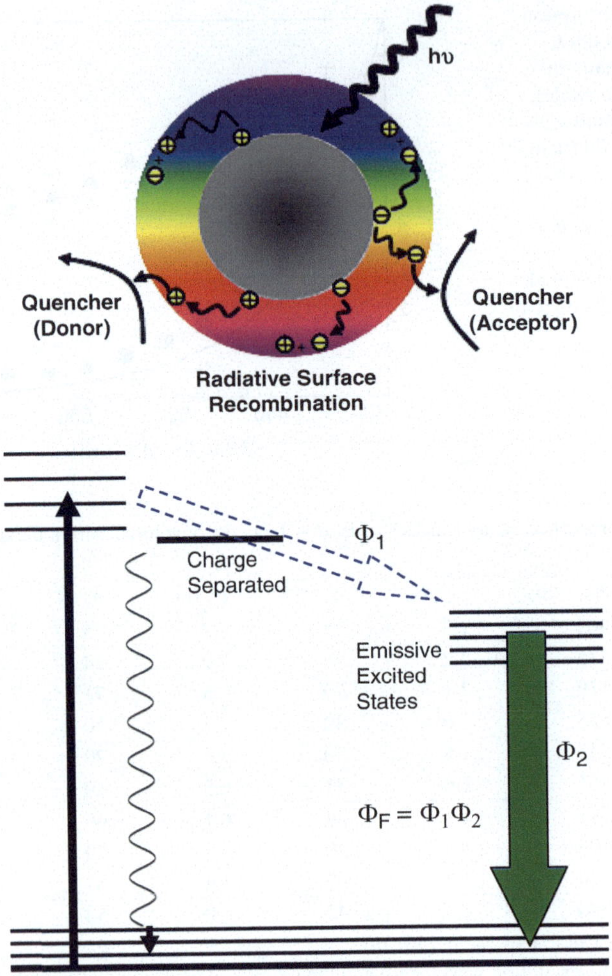

Fig. 4.7 A cartoon illustration (upper) and state diagram (lower) on the mechanistic framework for CDots

of diverse surface defect sites. Therefore, the "quantum-sized" for core carbon nanoparticles in CDots has different meanings and consequences from those in semiconductor nanocrystals with respect to the corresponding optical and electronic properties. Similarly, the size dependence in CDots, for which truly well-controlled experimental investigations are still needed, should also have different characteristics. In principle, a change in the carbon nanoparticle size alters the surface-to-volume ratio for the nanoparticle, which may change the surface defect sites, the radiative recombinations, and/or recombination paths, among other possible structural and energetic effects.

The defects in small carbon nanoparticles are understandably abundant, random in their distribution in an individual nanoparticle, varying from nanoparticle to nanoparticle, and different in structural details at different defect sites. The passivation of the defects via chemical functionalization of the small carbon nanoparticles in CDots is also subject to many variables, thus further exaggerating the diversity among the defect sites and also among different CDots. Yet, it is really stunning that the fluorescence emission properties, which according to the mechanistic framework should be dictated by the passivated defects, have exhibited such characteristic features consistently across different CDots. Most impressive have been the distinct patterns shared generally by different CDots in the excitation wavelength dependencies of fluorescence spectra (Fig. 4.5), quantum yields (Φ_F, Fig. 4.6), and lifetimes ($<\tau_F>$, Table 4.1 and Fig. 4.6).

On the two parameters Φ_F and $<\tau_F>$ describing the same fluorescence emissions (Fig. 4.6), a concern of the authors for (LeCroy et al. 2017b) was that the determination of each Φ_F value was based on the total fluorescence (the integration of the area under the observed fluorescence spectrum) while the measurement of fluorescence decay at each excitation wavelength covered only a portion of the emission spectrum, namely, a potential issue that the observed fluorescence emissions could have different average lifetimes at different emission wavelengths. The issue was examined by comparing fluorescence decays at the same excitation wavelength but different monitored emission wavelengths, and no meaningful difference was found in the observed decays, suggesting the homogeneity of the fluorescence spectrum with the same decay at different emission wavelengths. The elimination of such a concern makes the observed decoupling between Φ_F and $<\tau_F>$ in their excitation wavelength dependencies more significant mechanistically, because the decoupling (Fig. 4.6) translates to the pronounced excitation wavelength dependence of the $\Phi_F/<\tau_F>$ ratios (Fig. 4.8), which for molecular fluorophores typically express the radiative rate constants. In fact, the $\Phi_F/<\tau_F>$ ratios display rapid decreasing toward the red, apparently tracking the absorption spectral profile (Fig. 4.8) (LeCroy et al. 2017b).

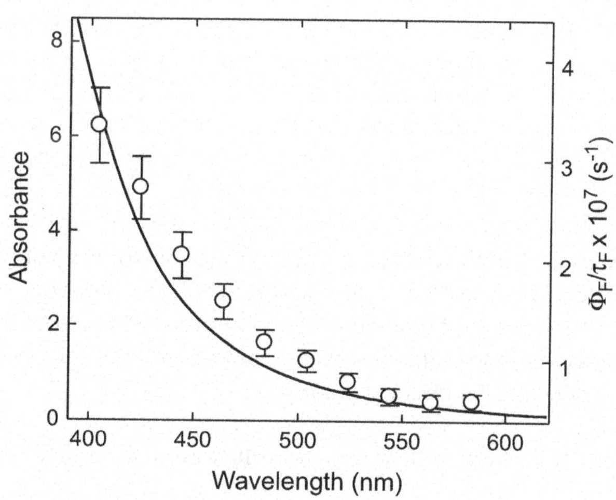

Fig. 4.8 The absorption spectrum (—) of EDA-CDots in aqueous solution and the $\Phi_F/<\tau_F>$ ratios for the different excitation wavelengths (O). (Originally part of Figure 5 in LeCroy et al. (2017b). Reproduced with permission)

The characteristic dependencies of fluorescence emission parameters (Figs. 4.6 and 4.8) subtend important information about the underlying photophysics of CDots. The mechanistic explanation and discussion on the observed excitation wavelength dependencies in (LeCroy et al. 2017b) were thorough and informative: "In CDots, the optical absorptions reflecting electronic transitions are associated with the core carbon nanoparticles, with likely multiple transitions in a distribution over the absorption wavelengths. Their absorption profile has been rationalized as being superposition from diverse state transitions, in which more strongly absorptive species exist in the blue region of the visible spectrum. It is therefore no surprise for the excitation wavelength dependent emissions in the EDA-CDots and other CDots reported in the literature. Mechanistically, the observed fluorescence properties may be explained in terms of two distinct photophysical processes and their efficiencies (Fig. 4.7): (i) the population of an emissive excited state (Φ_1), proceeding through spontaneous charge separation and trapping at surface sites, occurring after absorption, and (ii) the radiative recombination of charge carriers after trapping (Φ_2). The observed fluorescence quantum yield can then be viewed as a product of these two efficiencies such that $\Phi_F = \Phi_1\Phi_2$. Based on previous findings, the surface passivation in CDots mainly acts on the first term (Φ_1) in populating the emissive excited state, on which mechanistic reasons might include an improved stabilization of the trapping sites and/or a more effective electronic coupling between the core structure and surface moieties. The dramatic fluorescence enhancement effect produced on CDots by a proper passivation, together with the very large Φ_F values reported for many types of CDots, and the marked redox capabilities of photoexcited CDots strongly suggest that Φ_1 approaches unity, that is, no significant losses are expected in EDA-CDots during the formation of the emissive state. The latter process (Φ_2), as discussed herein, is primarily governed by radiative rate constants in competition with other deactivation pathways, specifically derived from inherent optical transitions in which the transition strengths are strongly correlated with brightness of fluorescent emissions." "The population of the emissive state from the earliest excited one (that is Φ_1) is likely driven by an excited-state reorganization expected to occur on the picosecond and sub-picosecond range, which can only be experimentally addressed by femtosecond spectroscopies, still very rare in the literature (Sciortino et al. 2017). Although it may be possible in principle that the excitation-dependence of Φ_F partially reflects changes of Φ_1, the simplest working assumption for the present purposes is the absence of losses during this step: (1) Φ_1 is excitation-independent and possibly close to unity, and (2) the excited-state relaxation and subsequent formation of the emissive state occurs with no significant decreases in the optical strength of the transition (namely, emission and absorption efficiencies), which would tend to lower the quantum yield." "In this scenario, the dependence of the quantum yield entirely arises from Φ_2 and for the same fluorescence emissions the observed quantum yield of the emitting state (Φ_F) is related to lifetime ($<\tau_F>$) by the equation,

$$\Phi_F \approx \Phi_2 = k_F \langle \tau_F \rangle \quad (4.1)$$

where k_F is the fluorescence radiative rate constant. Therefore, the $\Phi_F/<\tau_F>$ ratios plotted in Fig. 4.8 essentially express the radiative efficiency of the fluorescent transitions. While in molecular spectroscopy, the radiative rate constant k_F for a molecular fluorophore is generally excitation wavelength independent, Fig. 4.8 shows that this is not the case for CDots. Our data rather suggest that the optical transitions of CDots form a continuous distribution in which lower-energy emissive transitions are less and less allowed, as reflected by their progressively smaller radiative rates." "Furthermore, the radiative rate can be generally correlated with the electronic transition probabilities reflected by the corresponding optical absorption band,

$$k_F = 3\times10^{-9}\nu_0^2 \int \varepsilon d\nu \tag{4.2}$$

where ε is the extinction coefficient and ν_0 is the energy corresponding to the maximum transition energy (in wavenumbers). Therefore, the changes of the radiative rate are expected to reflect in corresponding changes of the absorption strength." "Indeed, across the investigated spectral range (a relatively narrow one in absolute terms), the term ν_0^2 in Eq. (4.2) changes slower than either the absorption coefficient or k_F, whose wavelength-dependence is very steep, and thus plays a relatively minor role. Considering this, a similarity between the radiative rate and the absorption profile (Fig. 4.8) suggests that the shape of the absorption spectrum largely reflects the different absorption strengths of the diverse transitions contributing to the overall spectrum. Although many transitions are involved and superimposed in the broad visible absorption, the absorption curve does not seem to be dictated only by the different abundance of CDots absorbing at different wavelengths, which would be the simplest assumption to make. On the contrary, the fundamental optical properties of CDots seem to be largely governed by the corresponding strengths of optical transitions, rather than inherent abundance. This can most clearly be illustrated by the tracing of the radiative rates, and, consequently, the fluorescence quantum yield of each occupied state." "While Eq. (4.2) cannot be strictly applied in a case where the optical absorption spectrum results from the superposition of many transitions, and therefore cannot be used to calculate the radiative rates, it still represents a framework to understand the photophysical relationship between the transition strengths and radiative rate constants. Equation (4.2) essentially states a proportionality between the radiative rate of an electronic transition and the allowedness of its optical absorption, hence the strength of the corresponding absorption band. Thus, the data in Figs. 4.5, 4.6, and 4.8 suggest that the optical absorption spectrum of CDots consists of an inhomogeneous distribution of transitions at different energies having different degrees of allowedness, systematically decreasing in the red, therefore producing emissions with similarly decreasing radiative rates and quantum yields" (LeCroy et al. 2017b).

An important takeaway from the investigations on the absorption and fluorescence emission properties of CDots may be such that the observed absorptions and fluorescence parameters are both associated with distributions of electronic transitions. However, the defects in small carbon nanoparticles are so abundant and diverse; their passivation via chemical functionalization might be forcing a "statistical convergence" in terms of confining (and narrowing) the distributions of electronic transitions to the specific patterns, which are consistent across different CDots. Such a strong "driving force" might explain the common experience that the basic fluorescence properties of CDots can readily be attained with some relatively simple and hardly stringent conditions for the surface passivation of the small carbon nanoparticles.

2.2 *Sample Inhomogeneity and Fractionation*

There is every expectation for a sample of CDots to be inhomogeneous due to variations in sizes of the core carbon nanoparticles and degrees of surface functionalization–passivation. The size distributions could be confirmed and quantified by transmission electron microscopy (TEM) and atomic force microscopy (AFM) analyses. Since the former is sensitive to the electron densities of the nanoscale entities, TEM images usually show sizes of the core carbon nanoparticles in the CDots. On the other hand, AFM height analyses may also include the "soft corona," with the estimated sizes often somewhat larger than those by TEM for the same CDots.

The sample inhomogeneity could be analyzed or exploited with the use of selected separation methods for not only an understanding of the sample inhomogeneity but also the isolation of CDots with more desirable properties such as significantly higher fluorescence performance. For example, the aqueous gel column originally developed for the separation of proteins and other biochemical species was employed for the fractionation of an as-prepared CDots sample (Wang et al. 2010a). Experimentally, the as-prepared sample of PEG_{1500N}-CDots (Fig. 4.1) was loaded onto an aqueous gel column packed with Sephadex™ G-100 gel for fractionation. With water as eluent, the fractions were collected and measured for their optical absorption spectra (Fig. 4.9). The excitation into the blue resulted in strong green fluorescence emissions. While the observed fluorescence spectra of different fractions were all rather similar (Fig. 4.9), their quantum yields were significantly different, progressively higher in later fractions, reaching Φ_F of 55–60% in the most fluorescent last fraction (Fig. 4.10).

The potential correlation of the observed different fluorescence quantum yields with sizes of the core carbon nanoparticles in the different fractions of CDots was first suspected but then dismissed on the basis of microscopy imaging results. The TEM images (Fig. 4.11) suggested no major differences between the pre-fractionation sample and the most fluorescent fraction, except for the latter being narrower in size distribution according to statistical analyses. These were generally supported by the AFM imaging results and the associated height analyses (Fig. 4.12). Instead, a likely correlation between the observed fluorescence properties and the degree (and effectiveness) of the surface passivation was proposed: "The changes in fluorescence quantum yields and lifetimes among different factions might be explained by varying degrees of surface passivation by PEG_{1500N}

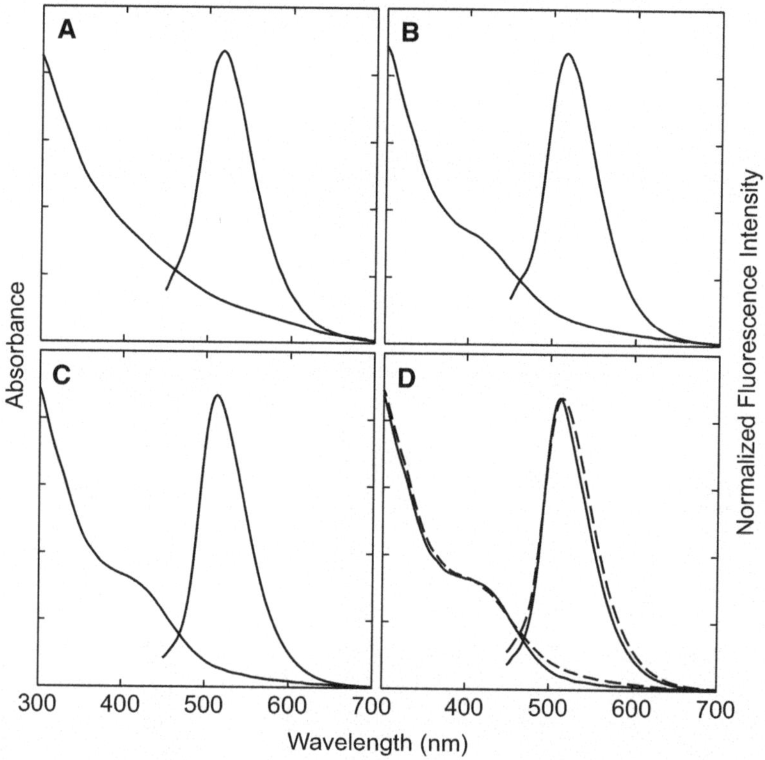

Fig. 4.9 Absorption and fluorescence (440 nm excitation) spectra of the fractions 1 (**a**), 3 (**b**), 5 (**c**), and the most fluorescent 7 (**d**). Also shown in (**d**) with dashed lines are the spectra of the as-prepared sample for comparison. (Originally Figure 1 in Wang et al. (2010a). Reproduced with permission)

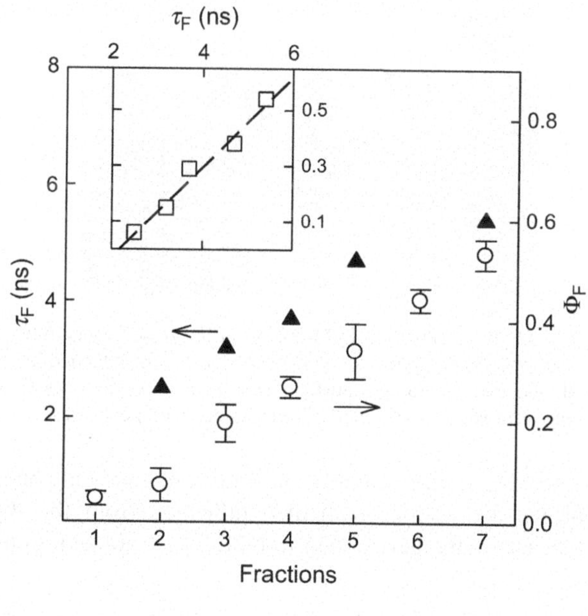

Fig. 4.10 Fluorescence quantum yields (○) and lifetimes (▲) for the different fractions of the PEG_{1500N}-CDots sample and the linear relationship between the observed yields and lifetimes (inset). (Originally Figure 2 in Wang et al. (2010a). Reproduced with permission)

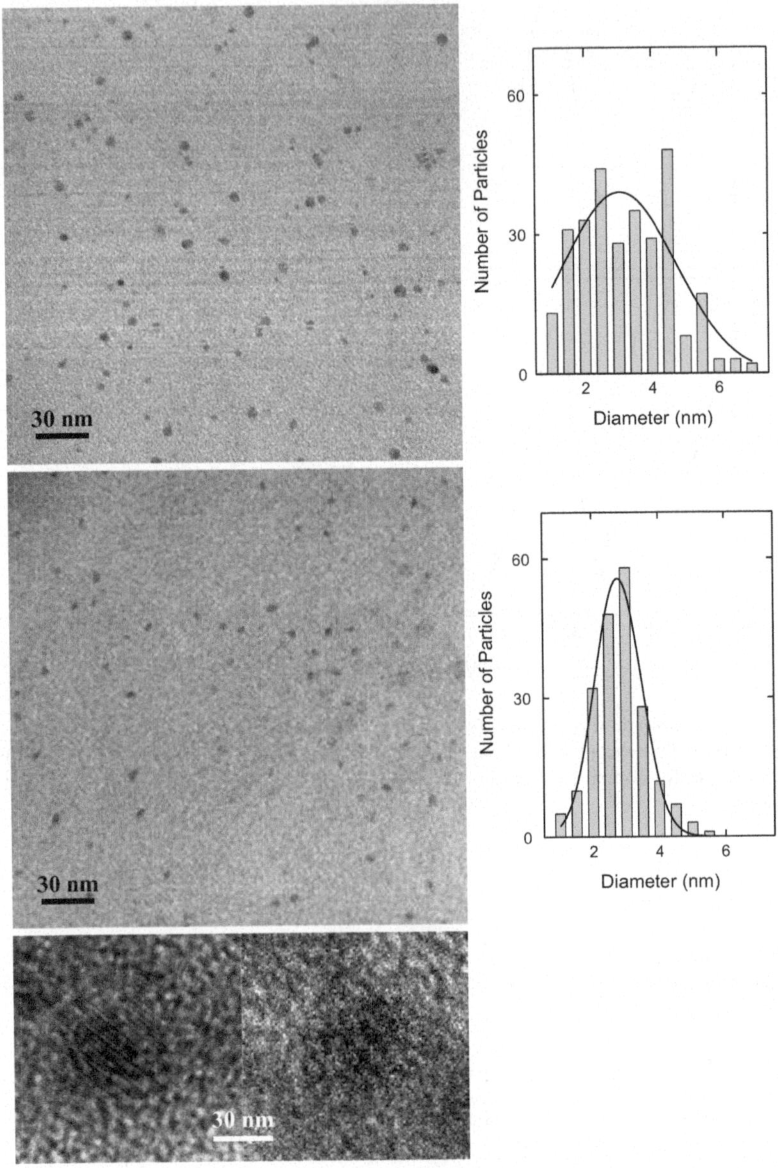

Fig. 4.11 Representative TEM images of PEG_{1500N}-CDots in the as-produced sample (upper) and in the most fluorescent fraction (lower, and also the attached high-resolution images of two dots), with the corresponding statistical size analysis results based on multiple images. (Originally Figure 3 in Wang et al. (2010a). Reproduced with permission)

molecules, both covalently via amide linkages and noncovalently in terms of strong surface adsorption, with also influence from the difference in particle sizes. Because on the gel column free PEG_{1500N} molecules eluted at last, a speculation is

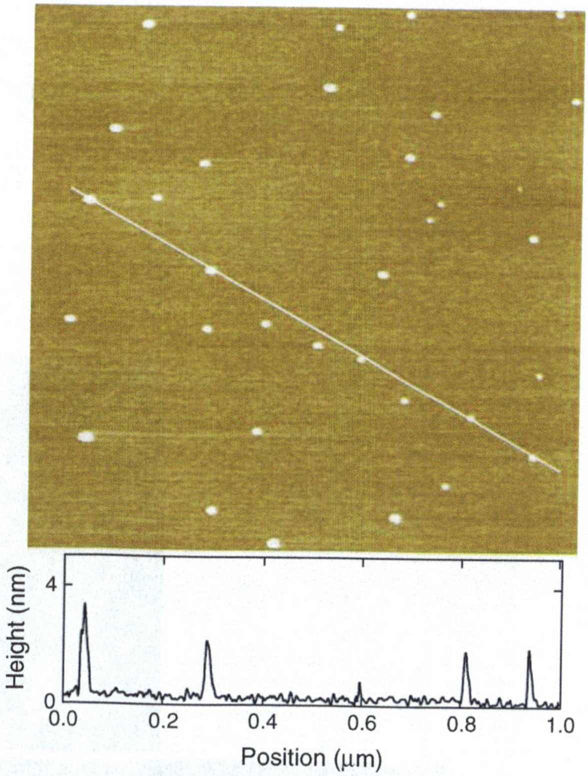

Fig. 4.12 AFM topography images of PEG_{1500N}-CDots in the most fluorescent fraction. (Originally Figure S1 in Wang et al. (2010a). Reproduced with permission)

such that the later fractions probably consisted of those CDots that were somewhat smaller in sizes and well-passivated with PEG_{1500N} molecules (thus making the dots behave a little closer to free PEG_{1500N} molecules)" (Wang et al. 2010a).

The authors of (Wang et al. 2010a) did not realize the other mechanistic implications of their reported results, such as the relevance to the discussion above based on (LeCroy et al. 2017b). Their observed widely varying fluorescence quantum yields among the different fractions were obviously not accompanied by any significant changes in fluorescence spectral profiles, suggesting that the more effective surface passivation could have major effects on the excited state processes in terms of higher efficiencies (yields), but not on the electronic transitions and their distributions. Such an understanding has been validated by the fluorescence properties obtained with fractions of other CDots (e.g., EDA-CDots (LeCroy et al. 2014)).

Since the work reported in (Wang et al. 2010a), the aqueous gel column fractionation has been established as a useful tool for the purification of CDots and the harvesting of better-performing fractions from the as-prepared dot samples. The proposed correlation of fluorescence brightness with the effectiveness in surface passivation of the small carbon nanoparticles in CDots has been confirmed in subsequent investigations on CDots of different surface functionalities.

2.3 *Fluorescence Performance*

CDots have been described generally as brightly fluorescent, but how bright has been a question. At the suggestion of a reviewer for (Wang et al. 2010a), an aqueous solution of the most fluorescent fraction of PEG_{1500N}-CDots (Wang et al. 2010a) and an ethanol solution of fluorescein (fluorescence quantum yield $\Phi_F = 0.79$) of matching optical density at 440 nm and 490 nm, respectively, were compared under different light conditions for visual appreciation. A Sony Cyber-Shot DSC-T20 digital camera was used for the photographs shown in Fig. 4.13.

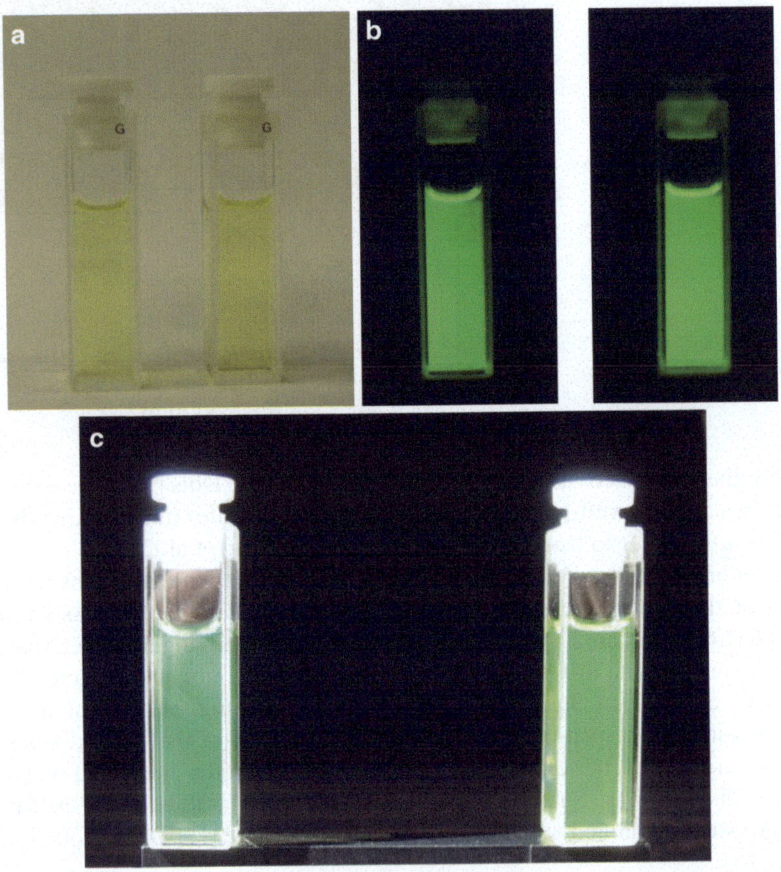

Fig. 4.13 (**a**) Photos under room light (left, aqueous PEG_{1500N}-CDots solution; right, fluorescein in ethanol solution. (**b**) Monochromated light for excitation and photographed through a 530 nm cut-off filter (left, aqueous PEG_{1500N}-CDots solution at 440 nm excitation; right, fluorescein in ethanol solution at 490 nm excitation). (**c**) Photos under sunlight (left, aqueous PEG_{1500N}-CDots solution; right, fluorescein in ethanol solution). (Originally Figures S2–S4 in Wang et al. (2010a). Reproduced with permission)

More quantitatively in terms of the observed fluorescence quantum yields, the most fluorescent were the best fraction of PEG_{1500N}-CDots (Φ_F up to 60% at 440 nm excitation) and the similarly aqueous gel column separated fraction of the CDots in which the core carbon nanoparticles were surface doped with ZnS and functionalized with PEG_{1500N} (denoted as C_{ZnS}-Dots, with Φ_F up to 78% at 440 nm excitation) (Anilkumar et al. 2011). It might also be argued that the fluorescence quantum yield measurement based on the relative method with the use of an organic dye as fluorescence standard would generally somewhat underestimate the Φ_F values of the CDots, especially for the excitation in the blue spectral region, because the observed absorption of CDots would always contain some light scattering contribution (versus none for the organic dye solution). Even without any corrections for the scattering effect, the available results suggest that CDots are competitive to the established CdSe/ZnS QDs for fluorescence performance in the comparable spectral region (Wang et al. 2010a; Anilkumar et al. 2011).

For individual dots as fluorescence probes, the brightness depends on both absorptivities at the excitation wavelengths and fluorescence quantum yields. The small carbon nanoparticles in CDots are strongly absorptive according to the estimated absorptivity values per molar concentration of carbons in the nanoparticles and also the observed large fluorescence radiative rate constants ($k_F \sim 10^8\ s^{-1}$). This coupled with the high fluorescence quantum yields makes CDots competitive fluorescence probes at the individual dot level, as demonstrated in the fluorescence microscopy imaging comparison with commercially supplied CdSe/ZnS QDs under "infinite dilution" (thus inter-dots distance larger than the light wavelength) conditions (Fig. 4.14) (Wang et al. 2010a).

A more interesting and practically significant demonstration on the performance of CDots as fluorescence probes was the comparison with commercially supplied CdSe/ZnS QDs for the fluorescence imaging in mice (Cao et al. 2012). The results again confirmed the competitive performance of CDots in vivo (Fig. 4.15).

3 Dot Structures, Excited State Processes, and Other Considerations

The structural details in the surface passivation layer of CDots (Fig. 1.1) have not been well understood, presenting challenges and opportunities for more systematic and quantitative investigations. The overwhelming majority of the reported CDots from the deliberate chemical functionalization synthesis have been based on the use of amine molecules for the functionalization under amidation reaction conditions. The results from various characterization methods have generally been consistent with the targeted functionalization but have not shown in an unambiguous fashion that the functionalization is indeed due predominantly to the formation of amide linkages. Nevertheless, the use of amine molecules for the functionalization in CDots has yielded more desirable outcomes in terms of

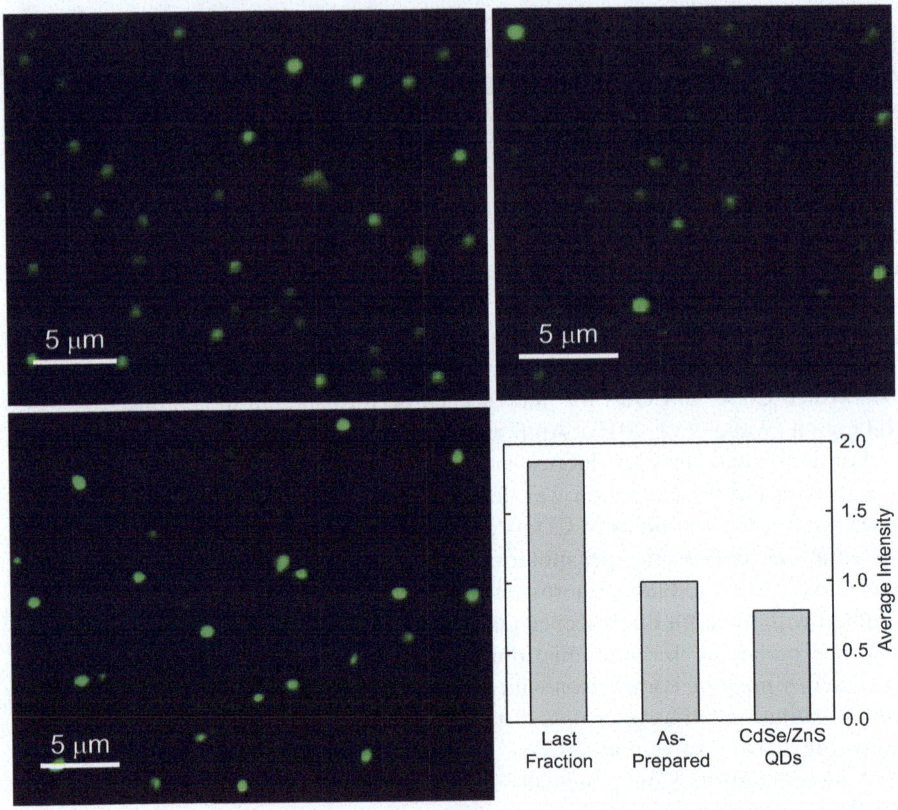

Fig. 4.14 Fluorescence microscopy images (458 nm excitation) of PEG_{1500N}-CDots in as-prepared sample (upper left) and in the most fluorescent fraction (lower left), and images of Invitrogen QD525PEG QDs (upper right). The bar–chart comparison was based on averaging 300 most fluorescent dots in each of the three samples. (Originally Figure 5 in Wang et al. (2010a). Reproduced with permission)

observed brighter fluorescence emissions. More quantitative mass balance analyses of the related CDots, such as those based on thermogravimetric analysis (TGA) measurements, have revealed that the amount of organic moieties on the dot surface is generally too high for just the amine molecules in amide linkages. Since the carbon nanoparticle surface-bound carboxylic acid groups must be limited in population, obviously impossible for every surface carbon to be oxidized into carboxylic acid in the processing of the carbon nanoparticles for simple reasons such as decarboxylation (readily for paired carboxylic acid groups), the number of amide linkages must be limited accordingly. Thus, there must be other bindings or interactions that are responsible for the higher than stoichiometrically possible population of amino moieties on the dot surface. More likely in the surface passivation layer of a CDot (Fig. 1.1) are the covalently bound amine molecules (in amide linkages) "supporting" the strong adsorptions of other amine molecules to the core carbon nanoparticle. Here the "supporting" could be by

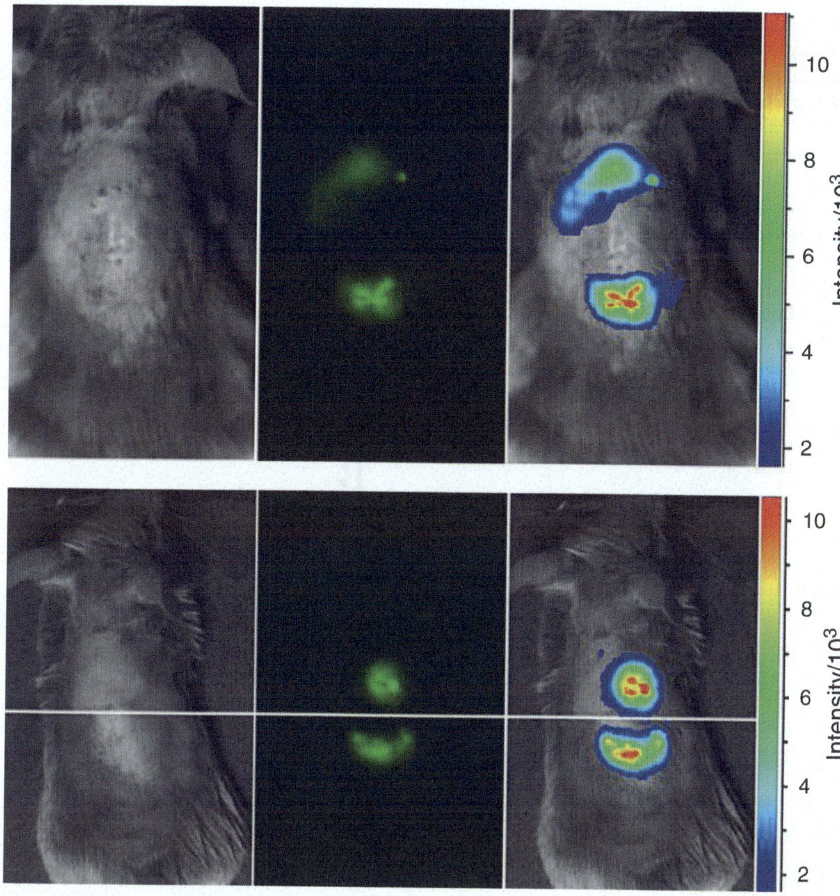

Fig. 4.15 Upper: subcutaneously injected PEG_{1500N}-CDots (upper spot) and C_{ZnS}-Dots (lower spot) of similar optical densities in mice (470 and 525 nm band-pass filters for excitation and emission, respectively). Lower: subcutaneously injected C_{ZnS}-Dots (upper spot) and Invitrogen Qdot® 525 ITK™ amino (PEG) QDs (lower spot) of similar optical densities in mice (434 nm band-pass filter for excitation and 474 nm cut-off filter for emission collection). The color-coded images (right) were processed with ImageJ from NIH. (Originally Figure 5 in Cao et al. (2012). Reproduced with permission)

physical entanglement. For example, PEG_{1500N} (a flexible long chain PEG molecule with primary amine groups at the two ends, Fig. 4.1) has been popular in the functionalization for CDots. The resulting CDots may each have a more compact surface passivation layer due to the entanglement of the flexible chains, which might be driven by the terminal amines all wanting to adsorb onto the carbon nanoparticle surface. In fact, it is known that the carbon nanotube surface is attractive to the adsorption of amine molecules (Fig. 4.16), with a rationale that the oxidative acid-treated carbon nanotubes are statically charged (Chattopadhyay et al. 2003; Li et al. 2004; Lu et al. 2011). Similarly, there may also be static

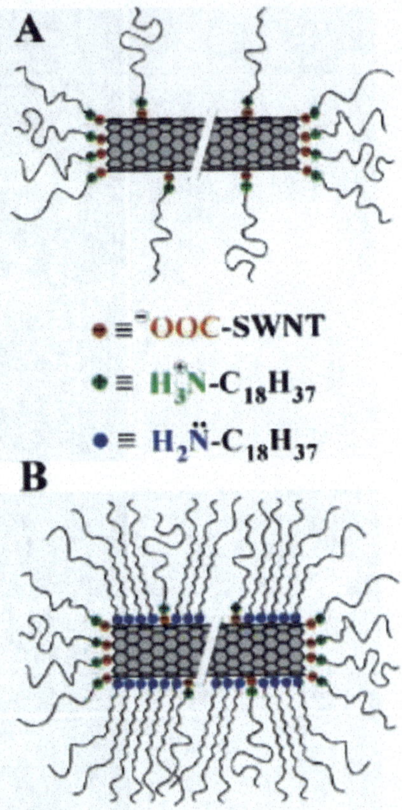

Fig. 4.16 Pictorial representation of possible octadecylamine (ODA) interactions with oxidative acid-purified SWNTs through (**a**) zwitterion formation; and (**b**) physisorption-assisted organization of ODA on SWNT sidewalls. (Reproduced from Chattopadhyay et al. (2003) with permission)

charges on the surface of small carbon nanoparticles, which could even be stabilized or enhanced by the covalent functionalization via amide linkages, thus enabling the adsorption of the amine molecules and consequently resulting in more effective surface passivation.

For an examination on the possible adsorption contributions to the surface passivation in CDots, Dr. Lingrong (Linna) Gu, who was a graduate student at the time in the Sun research group at Clemson University, spent many months looking into the correlation of repeatedly washing and dialyzing a PEG_{1500N}-CDots sample with the observed progressively decreasing fluorescence quantum yields, from about 20% in the starting sample to only 3–4% in the final sample that apparently contained CDots of much less amino moieties according to quantitative NMR analyses. That was in fact the first set of experimental evidence for the understanding that the fluorescence performance of CDots is sensitive to the effectiveness in the surface passivation of the core carbon nanoparticles. It also inspired various subsequent investigations on the de-functionalization of CDots, as well as the effort on "locking up" the surface passivation layer to prevent any unwanted de-functionalization.

The de-functionalization of CDots provides a view of the functionalization in the dot synthesis from the opposite angle, serving some valuable purposes such as confirming the targeted functionalization and examining and eliminating possible artifacts. It enables an unambiguous confirmation on the correlation between the fluorescence brightness and the effectiveness in the functionalization–passivation of core carbon nanoparticles in specific CDots. For example, EPA-CDots could be prepared by functionalizing the pre-processed and selected small carbon nanoparticles with 3-ethoxypropylamine (EPA, Fig. 4.1) under amidation reaction conditions (Yang et al. 2016). The observed optical absorptions of EPA-CDots were somewhat special, with a shoulder in the blue spectral region on top of the typical absorption spectrum of the core carbon nanoparticles (Fig. 4.17). The origin of the extra absorption feature was in question, and the gradual de-functionalization was used for an assessment on whether the absorption feature was genuine to the CDots.

The de-functionalization of EPA-CDots was accomplished by base-catalyzed hydrolysis under conditions for the EPA species being gradually striped from the dot surface. With the progression of the hydrolysis, the extra absorption shoulder decreased accordingly (Fig. 4.18), until in the end the absorption spectrum became essentially the same as that of the small carbon nanoparticles used in the synthesis of EPA-CDots. The conclusion was that the absorption shoulder must be genuine to

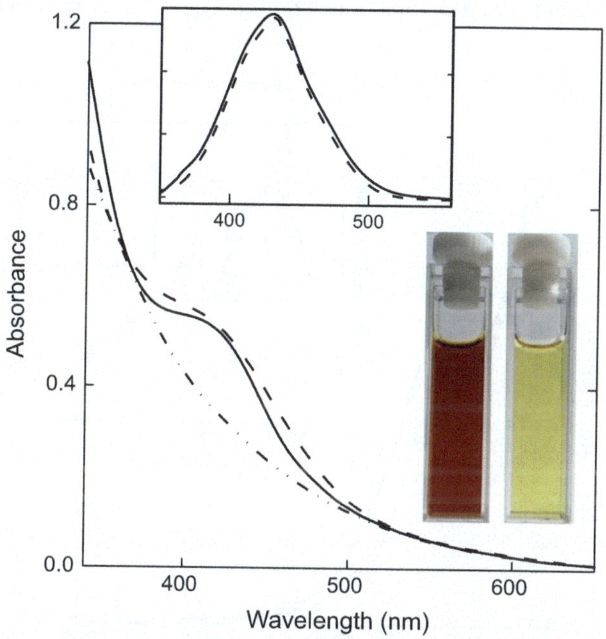

Fig. 4.17 The absorption spectrum of EPA-CDots in aqueous solution (——) is compared with those of the aqueous suspended carbon nanoparticles (-..-..-) and PEG_{1500N}-CDots (- - - -). The photos are for concentrated (left) and dilute (right) aqueous solutions of EPA-CDots. The other inset shows a comparison of two spectra corresponding to the extra absorption shoulder, obtained by subtracting the spectrum of the carbon nanoparticles (- - - -) or the final de-functionalized sample (——) from that of the EPA-CDots. (Originally Figure 2 in Yang et al. (2016). Reproduced with permission)

the CDots, not associated with some "impurities" introduced in the dot synthesis. However, the results have prompted a more difficult and important question on whether the simple EPA functionalization could actually alter the electronic transitions known to be associated with π-plasmons in the small carbon nanoparticles. The description and discussion quoted below on the observed fluorescence properties might be helpful, but the question still remains.

Per (Yang et al. 2016), "Fluorescence spectra of the same samples collected at different de-functionalization reaction points are shown in Fig. 4.18. The spectral profiles are similar between different samples, but the emission intensities are progressively lower, so are the observed fluorescence quantum yields (Fig. 4.18). However, the decrease in yields at 440 nm excitation is relatively more rapid than that at 400 nm excitation, likely due to the fact that upon de-functionalization the absorption shoulder responsible for the enhanced fluorescence emissions in the original EPA-CDots disappears quickly (Fig. 4.18)." "The samples with different degrees of de-functionalization were also evaluated by fluorescence decay measurements. As compared in Fig. 4.19, there are no major changes in the fluorescence decay curves between the samples, suggesting that the dynamic processes in the emissive excited states are decoupled from the dramatic decreases in fluorescence quantum yields upon de-functionalization (Fig. 4.18)" (Yang et al. 2016).

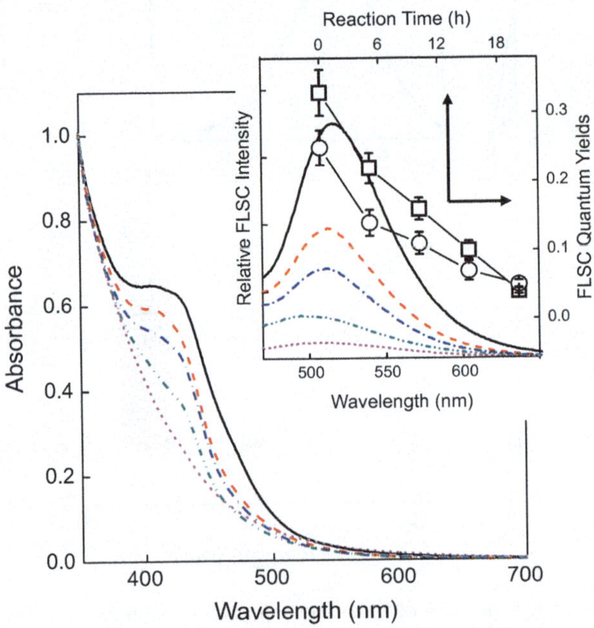

Fig. 4.18 Absorption and fluorescence (FLSC, 440 excitation, in the inset) spectra of EPA-CDots (——) and the four samples with different degrees of de-functionalization (D_1: — —, D_2: -.-.-, D_3: -..-..-, and D_4: - - -). Also shown in the inset are the changes in fluorescence quantum yield with the de-functionalization at 400 nm (○) and 440 nm (□) excitations. (Originally Figure 8 in Yang et al. (2016). Reproduced with permission)

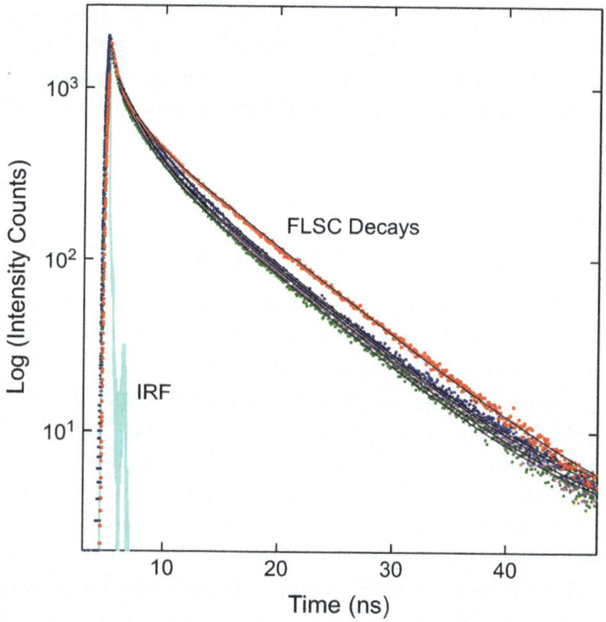

Fig. 4.19 Observed fluorescence decays of EPA-CDots (red) and the four samples with different degrees of de-functionalization (D_1: blue, D_2: magenta, D_3: olive, and D_4: navy), all at 440 nm excitation and the monitored emission centered at 510 nm. The bi-exponential fitting curves for all of the decays are also shown (black). (Originally Figure 5 in Yang et al. (2016). Reproduced with permission)

Independent of the difficult and important question discussed above on the extra absorption shoulder of EPA-CDots, the results from the de-functionalization study support strongly the correlation between fluorescence brightness and passivation effectiveness. In addition, the obvious decoupling between fluorescence quantum yields and lifetimes of the EPA-CDots samples with different degrees of de-functionalization is rather similar to the decoupling observed in the study of EDA-CDots (Fig. 4.6) (LeCroy et al. 2017b). They likely share the same mechanistic origins (Fig. 4.7).

Likely also sharing the same mechanism is the significantly enhanced fluorescence brightness (higher quantum yields) for CDots hosted in a more confined environment, such as the embedding of CDots in a polymer matrix, completely decoupled from the absence of any changes in fluorescence decays (Liu et al. 2016).

The apparent contribution of adsorption to the surface passivation of carbon nanoparticles in CDots presents some vulnerability with respect to de-functionalization under some application conditions or in storage over time. One strategy explored to address the issue was to use established crosslinking agents to lock up the surface passivation layer in CDots, thus to stop any unwanted de-functionalization. However, the implementation of such a strategy was found to be more challenging than anticipated. The difficulty was due to interferences such as the inter-dot crosslinking competing with the desired intra-dot crosslinking. Nevertheless, the former was later exploited successfully for the preparation of fluorescence probes with multiple crosslinked CDots for much enhanced fluorescence performance at the individual probe level (Anilkumar et al. 2013), as discussed in more detail in Chap. 6.

4 Redox Characteristics and Mechanistic Implications

The redox characteristics of CDots in the ground state have not been explored in any systematic fashion. Since carbon nanoparticles are considered electron rich, including their popular uses in reductive processes as a major part of the commercially available catalysts (Pd/C, for example) for various organic reduction reactions, one may expect CDots to be similar. This is obviously very different from fullerenes, which are well known as excellent electron acceptors even in the ground state, including the formation of ground-state charge transfer complexes with some electron donor molecules. Also considerably different from fullerenes is that photoexcited CDots are both potent electron donors and acceptors. These major differences support the argument that fullerenes, while zero-dimension in shape, are really large molecules of distinctly defined electronic structures with the characteristic electron-deficient nature. CDots derived from small carbon nanoparticles, on the other hand, are truly nanomaterials at the zero-dimension capable of manifesting both electron-rich and electron-deficient characteristics in the same nanostructure.

4.1 *Fluorescence Quenching via Charge Transfers*

Back in 2009 (Wang et al. 2009), it was found that the fluorescence emissions of PEG_{1500N}-CDots (Fig. 4.1) could be quenched efficiently by both classical electron donor and acceptor molecules, *N,N*-diethylaniline (DEA, 0.88 V versus NHE) and 2,4-dinitrotoluene (DNT, −0.9 V versus NHE), respectively. These quenchers are colorless, thus no interference to the excitation of CDots in the visible spectral region. The observed Stern–Volmer constant ($K_{SV} = \tau_F{}^{\circ} k_q$) of 83 M^{-1} for the DNT quenching of fluorescence intensities (quantum yields) was unusually large, as it would correspond to the bimolecular rate constant k_q of 2.1×10^{10} M^{-1} s^{-1} (based on the estimated average lifetime $\tau_F{}^{\circ}$ around 4 ns), thus beyond the diffusion control in a toluene solution. However, in the same toluene solution, the quenching of fluorescence lifetimes was close to normal, with K_{SV} of 32 M^{-1} and k_q of 8×10^9 M^{-1} s^{-1}, still within the norm of diffusion control.

The quenching by DEA was both simpler and more complicated, but generally less efficient than the quenching by DNT (Wang et al. 2009). On the simple side, the fluorescence intensities and lifetimes of PEG_{1500N}-CDots in methanol were quenched almost equally, with the estimated K_{SV} of 19 M^{-1} and k_q of 4.8×10^9 M^{-1} s^{-1}. More complicated were the downward-curved Stern–Volmer plot at higher DEA concentrations and the observed strong solvent dependence, with the quenching much less efficient in the less polar solvent chloroform (Fig. 4.20). For a significantly weaker electron donor such as diethylamine (1.55 V versus NHE), the quenching was also considerably less efficient, with the estimated K_{SV} of only about 0.3 M^{-1}.

At the time, these fluorescence quenching results contributed to the proposed mechanistic framework for CDots, in which the observed fluorescence emissions were attributed to radiative recombinations of the electrons and holes trapped at the passivated surface defect sites (Fig. 4.7). Qualitatively the fluorescence quenching by both electron donor and acceptor molecules could be rationalized as being due to their interrupting the radiative recombinations via interactions with the trapped holes and electrons, respectively (Fig. 4.7). However, such a simple rationale was not sufficient to explain many of the other details in the results, such as the extraordinarily efficient DNT quenching for fluorescence quantum yields, but much less so for fluorescence lifetimes.

In a more recent revisit of the fluorescence quenching with DNT (LeCroy et al. 2017a), the sample of structurally more compact EDA-CDots (Fig. 4.1) was used. The quenching was studied systematically by using both steady-state and time-resolved fluorescence techniques at different excitation wavelengths. The quenching parameters (K_{SV} and k_q values) in methanol were generally similar to those obtained previously for PEG_{1500N}-CDots in toluene at the same excitation wave-

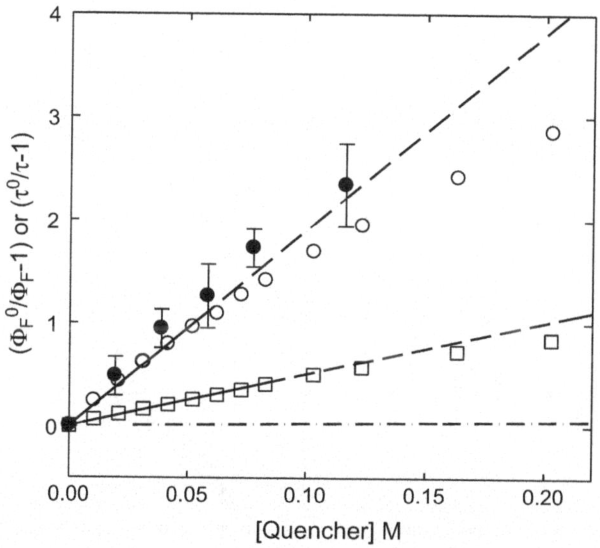

Fig. 4.20 Stern–Volmer plots for the quenching of fluorescence quantum yields (400 nm excitation) of PEG_{1500N}-CDots by *N,N*-diethylaniline (DEA) in methanol (○) and chloroform (□) and for the quenching of fluorescence lifetimes (407 nm excitation) in methanol (●). The low-concentration portion of the same plot for diethylamine as the quencher in methanol (-.-.-) is also shown for comparison. (Originally Figure 4 in Wang et al. (2009). Reproduced with permission)

length (425 nm). For excitation at longer wavelengths, the Stern–Volmer plots similarly downward curved at higher DNT concentrations (Fig. 4.21, compared to Fig. 4.20). It was also shown unambiguously that fluorescence spectral profiles did not change at all with the quenching (Fig. 4.22), suggesting again the homogeneity of the underlying emissions. The same conclusion for the DEA quenching based on the absence of any changes in fluorescence spectral profiles with DEA concentrations was also reached in a separated study (LeCroy et al. 2017b).

On the extraordinarily efficient (beyond the diffusion controlled limit) DNT quenching of fluorescence quantum yields, decoupled from that of lifetimes, static quenching contribution became a logical explanation (LeCroy et al. 2017a): "Static quenching is in many cases associated with the formation of excited state complexes, those between the emissive species and the quencher molecules. In this case, however, there is no experimental evidence for the formation of such complexes, with no changes in the observed absorption and fluorescence spectra of the dot solutions at all of the quencher concentrations used." It may also be argued that the decoupling between the quenchings of fluorescence quantum yields and lifetimes should be understood in the mechanistic framework of roughly two excited state processes, Φ_1 and Φ_2 (Fig. 4.7), on different time scales. The near-neighbor static quenching contribution in the quenching of fluorescence quantum yields could affect both processes, while the dynamic quenching of fluorescence lifetimes could only speed up the decays of the emissive excited states (the slower Φ_2 process in Fig. 4.7).

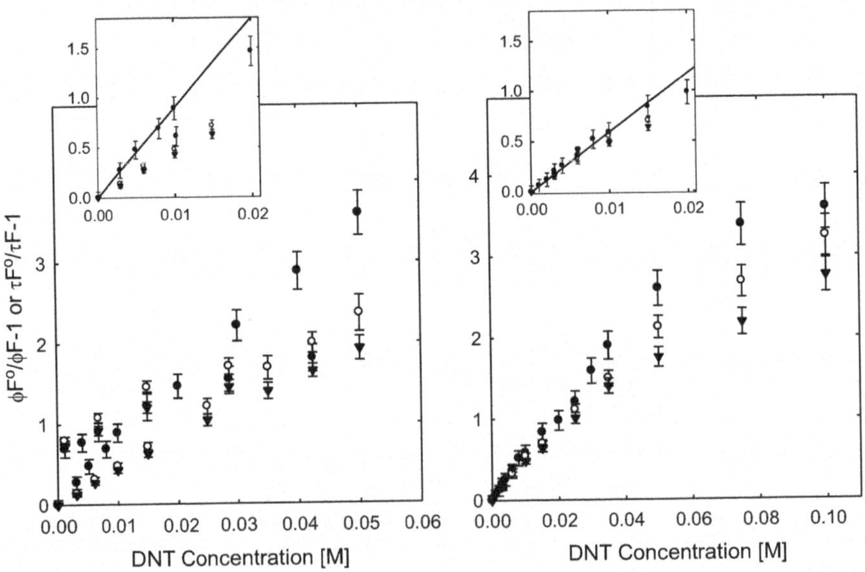

Fig. 4.21 Stern–Volmer plots for the quenching of fluorescence intensities (●) and average fluorescence lifetimes (with emissions monitored at the spectral peak, ○, and at 600 nm ▼) of EDA-CDots in methanol solutions with different concentrations of 2,4-dinitrotoluene (DNT) and excitations at 425 nm (left) and 475 nm (right). (Originally parts of Figure 4 in LeCroy et al. (2017a). Reproduced with permission)

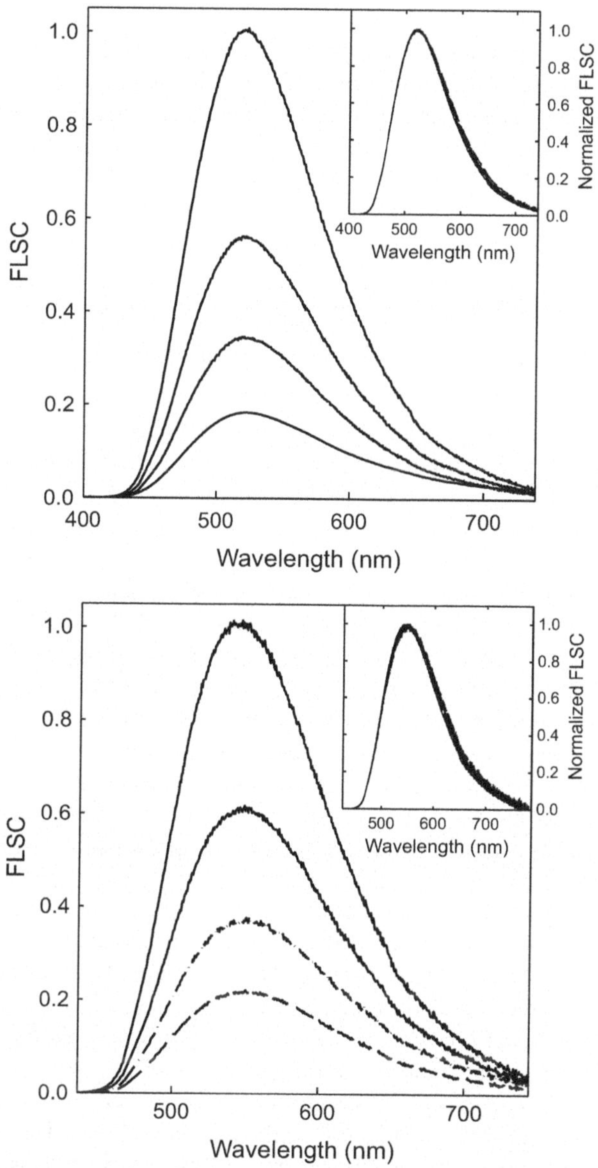

Fig. 4.22 The quenching of fluorescence (FLSC) intensities of EDA-CDots by 2,4-dinitrotoluene (DNT) in methanol solution at 425 nm (upper) and 475 nm (lower) excitations. In each figure the spectra with decreasing intensities correspond to quencher concentrations of 0–0.05 M for the 425 nm excitation (upper) and 0 to 0.1 M for the 475 nm excitation (lower). Shown in the insets are the corresponding normalized spectra. (Originally parts of Figure 3 in LeCroy et al. (2017a). Reproduced with permission)

The main takeaways from the observed fluorescence quenching behaviors are the following: (1) the photoinduced redox characteristics of CDots are unique, at least in comparison with those of fullerenes, capable of serving as both potent electron donors and acceptors; (2) CDots in the photoexcited states may generally be stronger electron donors than acceptors; (3) static quenching contributions could be substantial; and (4) the fluorescence quenching may serve as a window into which the probing by using additional tools such as ultrafast optical spectroscopy techniques will prove rewarding. Mechanistically in the presently adopted framework for CDots, the quenching (in a simplistic way of thinking) could be targeting the charge separated species to interrupt their radiative recombinations and/or the resulting emissive excited states, which should correspond to different time scales. The relative contributions of the two excited state processes in the experimental quenching results remain to be understood in a more quantitative fashion. Practically, however, the highly efficient quenching, including the substantial static component in the quenching of fluorescence intensities, has made CDots extremely popular fluorescence sensors.

4.2 Fluorescence Sensing of Metal Ions

A long list of metal ions can quench fluorescence emissions of CDots via mostly charge transfer interactions, which have naturally made CDots widely pursued fluorescence sensors for these metal ions (Gao et al. 2016). Beyond the coverage of most reviews on the topic, one extraordinary finding in investigations on the sensing of metal ions based on fluorescence quenching with CDots has been the ultra-large "apparent" Stern–Volmer quenching constants (K_{SV}), with some reported K_{SV} values of more than 100,000 M^{-1} or even 500,000 M^{-1}. Here the characterization of these K_{SV} numbers as "apparent" is for the fact that the Stern–Volmer equation has become mechanistically invalid in the treatment of such extraordinary fluorescence quenching results. These astronomically large slopes in the empirical linear quenching plots obviously make the sensors extremely sensitive practically, but without the necessary backing based on mechanistic understandings, there could be major issues with the predictability and reproducibility, among others, in the design, preparation, and uses of the CDots-derived sensors.

There is no magic in fluorescence quenching. The extraordinarily large slopes found in the empirical linear quenching plots, on which the reported sensors have been based, could only be due to an equally extraordinary underestimate of the metal ion concentrations actually experienced by the photoexcited CDots, namely, the "local concentrations" of the metal ions must be considerably larger than the bulk concentrations. Simplistically these may be considered in the framework of static quenching effects commonly discussed in photophysics books. Even so, the fluorescence quenching behavior with CDots-derived sensors found in many reported studies is still far beyond the norm of "classical" static quenching, suggesting extraordinarily strong driving forces or conditions for the ultrahigh local analyte concentrations. Interestingly, the behavior is apparently analyte dependent, mostly in the sensing of some metal cations, and also varying among different CDots samples. As discussed earlier, the surface of CDots may be statically charged, which may facilitate the effective adsorption of the metal ions for essentially fluorescence quenching within the dot sphere. In fact, most of the dot samples in the reported sensing studies were prepared by the carbonization of organic precursors, with the resulting dot structures porous like sponges (see more detailed discussion in Chap. 5). Therefore, the fluorescence quenching by the metal ions soaked in the dot structures should be considered as effectively intra-dot in nature.

4.3 *Dot Surface Doping/Coating with Metals*

The extremely efficient intra-dot or close-contact fluorescence quenching via charge transfer interactions has been demonstrated in the surface doping of CDots with a noble metal (Xu et al. 2012). Experimentally, an aqueous solution of PEG_{1500N}-CDots was first examined for photostability. Both optical absorption and fluorescence emissions of the sample were found stable with respect to photoirradiation with a 450 W xenon source coupled with filters for the 425–720 nm spectral range for up to 10 h. The observed fluorescence spectral profiles and intensities were essentially unchanged (Fig. 4.23), suggesting that there were no meaningful photochemical reactions in the dot solution in the absence of any other substances. In the presence of a noble metal salt, however, specifically the platinum(IV) compound H_2PtCl_6 or gold(III) compound $HAuCl_4$ in the aqueous solution, the same photoirradiation resulted in dramatic decreases in observed fluorescence intensities over just a short period of time (Fig. 4.23), namely, substantial quenching of the fluorescence emissions. In fact, at the Pt(IV) salt concentration of 0.3 mM, the photoirra-

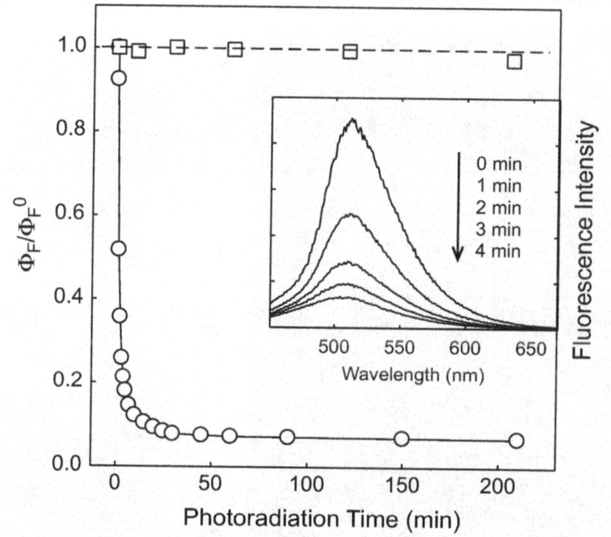

Fig. 4.23 Observed fluorescence quantum yields (Φ_F, 440 nm excitation, normalized against the value at time zero Φ_F^0) of PEG_{1500N}-CDots in aqueous solution without (□) and with the Pt(IV) compound (O) after visible light photoirradiation for various periods of time. The corresponding spectral intensity changes for the first few (O) data points (with increasing photoirradiation times in the arrow direction) are shown in the inset. (Originally Figure 2 in Xu et al. (2012). Reproduced with permission)

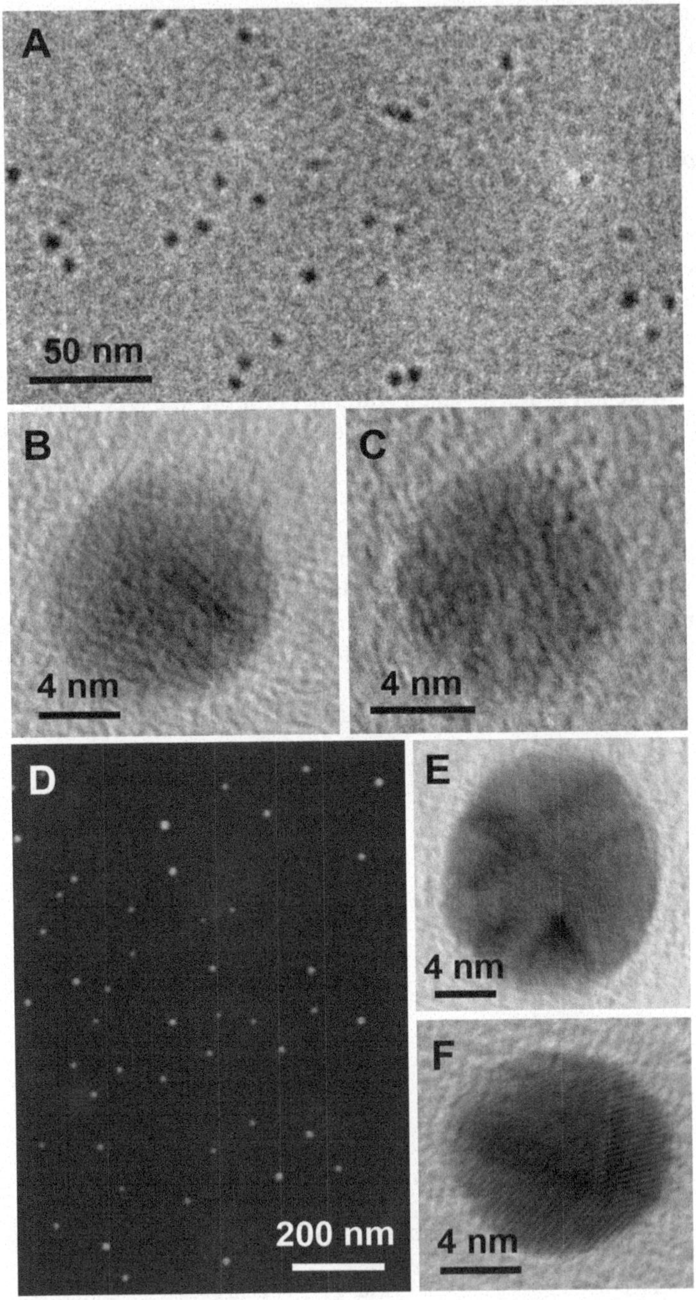

Fig. 4.24 TEM results on PEG_{1500N}-CDots with the photo-deposited platinum metal (**a**, **b** and **c** at high resolution) or gold metal (**d** in the Z-contrast mode, and **e** and **f** at high resolution). In **f**, the spacing in the crystal lattice fringes of 0.29 nm is consistent with fcc gold (110). Except for **d** (a dark-field image), all others are bright-field images. (Originally Figure 6 in Xu et al. (2012). Reproduced with permission)

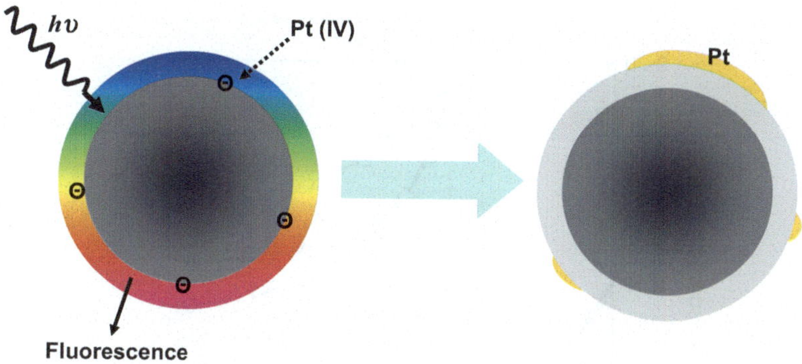

Fig. 4.25 A schematic illustration on the photoreductive deposition of Pt metal onto the CDot surface and the subsequent fluorescence quenching. The rainbow "shell" is to indicate that the fluorescence is likely from the passivated carbon nanoparticle surface. (Originally Figure 5 in Xu et al. (2012). Reproduced with permission)

diation for only 5 min resulted in the quenching of fluorescence intensities by close to 90% (Xu et al. 2012). The dramatic fluorescence quenching was not due to the Pt(IV) salt in the dot solution according to control experiments. Rather, it was the work of Pt metal on the dot surface from the photoreductive deposition (Fig. 4.24). The photoexcited CDots acted as potent electron donors for the reduction of Pt(IV) into Pt metal on the dot surface. Then, as illustrated in Fig. 4.25 on the Pt-doped CDots, the Pt metal, known for high electron affinity, must be taking the trapped electrons produced in the photoinduced charge separation, thus interrupting radiative recombinations and diminishing the fluorescence emissions. The effect of the dot surface-doped noble metal "concentrating" electrons has been confirmed in the use of CDots as photocatalysts (Cao et al. 2011a) (see more details in Chap. 6).

Unlike platinum, gold has plasmon absorption in the visible spectrum, allowing the monitoring of Au metal formed in the photoirradiation (Fig. 4.26). The Au metal doped on the surface of CDots could also be probed by transmission electron microscopy (TEM) imaging (Fig. 4.24).

Fluorescence decays of the CDots were clearly not affected by the surface-doped Pt or Au metal (Fig. 4.27), again completely decoupled from the dramatic quenching of fluorescence intensities. The static nature for the latter is obviously consistent with the essentially intra-dot or close-contact arrangement for the quencher, namely, the noble metal doped on the dot surface. Interestingly, however, the fluorescence spectral changes with the quenching (relatively minor, but still readily observed) were different from the no change at all found in other quenching studies, such as organic electron donor DEA and acceptor DNT as quenchers discussed above (LeCroy et al. 2017a). Mechanistic implications of the difference remain to be explored.

The reduction of metal cations by photoexcited CDots in solution represents an interesting and mechanistically significant example for the unique photoinduced redox characteristics of CDots. Beyond that, the metal-doped CDots thus produced are valuable to other studies and/or applications, such as photocatalysis, photosensitization, and optoelectronics (see more details in Chap. 6).

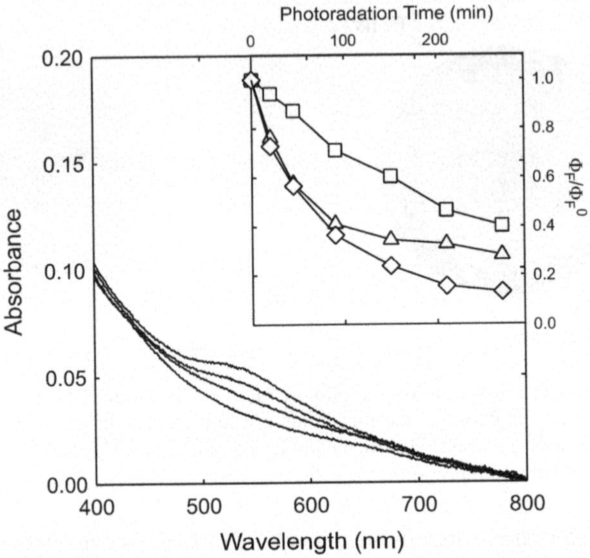

Fig. 4.26 The emergence and increases of the gold plasmon absorption band (around 550 nm) after aqueous solution of PEG_{1500N}-CDots with the Au(III) compound were photoirradiated for 15, 270, 330, and 390 min. Shown in the inset are the normalized fluorescence quantum yields (Φ_F/Φ_F^0 with excitation at 400 nm, □; 440 nm, ◇; and 500 nm, Δ) of the same solution after visible light photoirradiation for various periods of time. (Originally Figure 4 in Xu et al. (2012). Reproduced with permission)

5 Other Surface Passivation Schemes

Amidation of the surface-bound carboxylic acid groups has been used extensively in the functionalization of small carbon nanoparticles with amine molecules or polymers, but it is hardly an exclusive option for the desired surface passivation. The pre-processed and selected small carbon nanoparticles could also be functionalized by amines and other molecules in thermally induced reactions. While not so selective in terms of the targeted binding patterns or structures, these reactions have yielded dot samples of fluorescence properties comparable to those of the CDots from the classical amidation synthesis. For example, the functionalization of small carbon nanoparticles with oligomeric polyethylenimine (PEI) was achieved by microwave-assisted thermal processing for PEI-CDots (Hu et al. 2016; Ge et al. 2018). As compared in Fig. 4.28, the absorption and fluorescence spectral and quantum yield results of PEI-CDots are rather similar to those of EDA-CDots (prepared via amidation (LeCroy et al. 2014, 2017b)). The photoinduced redox characteristics of PEI-CDots were also found to be quite similar to those of EDA-CDots in terms of their largely the same behaviors in the fluorescence quenching with electron donor (DEA) and acceptor (DNT) quenchers (Ge et al. 2018).

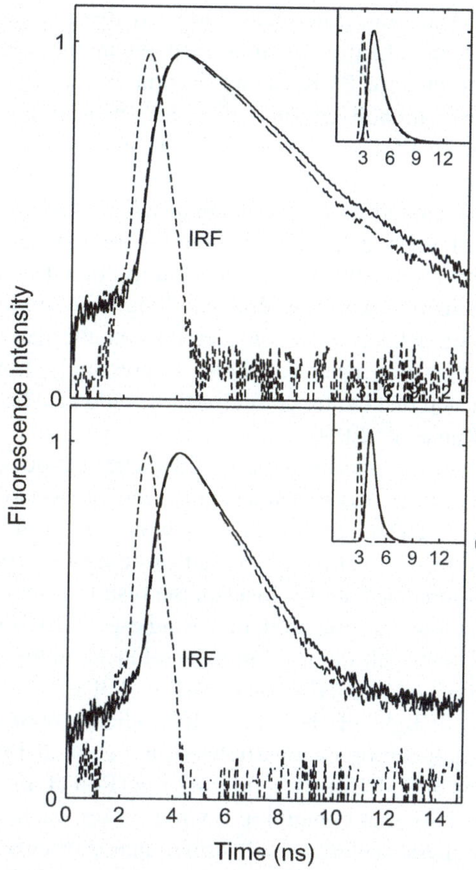

Fig. 4.27 Observed fluorescence decays (log scale for the fluorescence intensities, excited at 407 nm, and monitored with 470 nm narrow band-pass filter) of PEG_{1500N}-CDots in aqueous solution without (——) and with (— —) the photo-deposited platinum metal (upper) or gold metal (lower). The same data are plotted on linear scale in the insets. (Originally Figure 8 in Xu et al. (2012). Reproduced with permission)

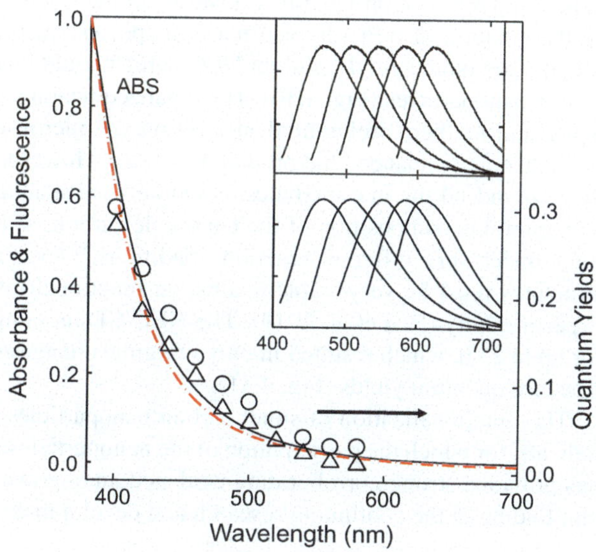

Fig. 4.28 Absorption spectra (ABS) and fluorescence quantum yields at different excitation wavelengths for EDA-CDots (——, ○) and PEI-CDots (-.-.-, Δ). Inset, normalized fluorescence spectra of EDA-CDots (upper) and PEI-CDots (lower) corresponding to excitations (from left to right) at 400, 440, 480, 520, and 560 nm. (Originally Figure 7 in Ge et al. (2018). Reproduced with permission)

The same microwave-assisted thermal processing approach was used for the functionalization of small carbon nanoparticles with poly(*N*-vinylcarbazole) (PVK) polymer for PVK-CDots (Rednic et al. 2015). Such a functionalization approach was rationalized such that the absorption of microwave energy by the carbon nanoparticles would make their surface very hot, causing partial carbonization of the organic species, namely, that a part of the PEI molecule or PVK polymer would be "melted onto" the nanoparticle surface for the attachment, with the remaining part serving the passivation function. Consistent with such a rationale was the observed stability of the dots thus prepared against unintentional de-functionalization in solvent washing, dialysis, long-term storage, etc. In fact, this functionalization approach has been considered as an alternative strategy in "locking up" the surface passivation layer in CDots. Interestingly, however, more recent results have suggested that the functionalization could also be due at least in part to radical additions (Ren et al. 2019).

In the microwave-assisted thermal processing for the functionalization of preprocessed and selected small carbon nanoparticles with *N*-ethylcarbazole (NEC) for NEC-CDots, highly unlikely would be for the small and structurally compact NEC molecule to allow the partial carbonization for the kind of functionalization scheme rationalized above. Instead, the results from the characterization of the NEC-CDots sample thus prepared, including especially NMR results (Fig. 4.29), were more consistent with the formation of adducts, namely, the carbon nanoparticles covalently attached with NEC species (Fig. 4.30).

In light of the successful radical copolymerization of *N*-vinylcarbazole with small carbon nanoparticles (Yang et al. 2018b), a radical addition reaction scheme for NEC-CDots was proposed in (Ren et al. 2019): "Molecules structurally similar to NEC but containing a vinyl group, such as *N*-vinylcarbazole, could be copolymerized with the small carbon nanoparticles in thermally induced radical polymerization reactions. Without a vinyl group NEC is obviously incapable of any classical radical polymerization reactions. Instead, in the microwave treatment there might be the creation of NEC-derived reactive species, such as radicals from losing an α-hydrogen on the ethyl group in NEC, which could bind to surface carbons of the carbon nanoparticles (Fig. 4.30). The localized heating around the carbon nanoparticles due to their preferential absorption of microwave irradiation might have facilitated and enhanced the binding reactions. However, the amount of radicals, if they are indeed the intermediates, is unlikely substantial and sufficient to result in meaningful decomposition of the nanoscale carbon, as found in other radical reactions under very different reaction conditions." The radical addition-derived dot structures must be very stable against de-functionalization, which was confirmed experimentally (Ren et al. 2019). The NEC-CDots sample was refluxed in toluene for up to 25 h, which resulted in only marginal changes/decreases in observed fluorescence quantum yields (Fig. 4.31).

The functionalization of small carbon nanoparticles via the addition of organic radicals, for which the participation of the nanoparticles in radical copolymerization has provided strong corroborating evidence, may prove to be an extremely important finding in the continuing research and development of CDots.

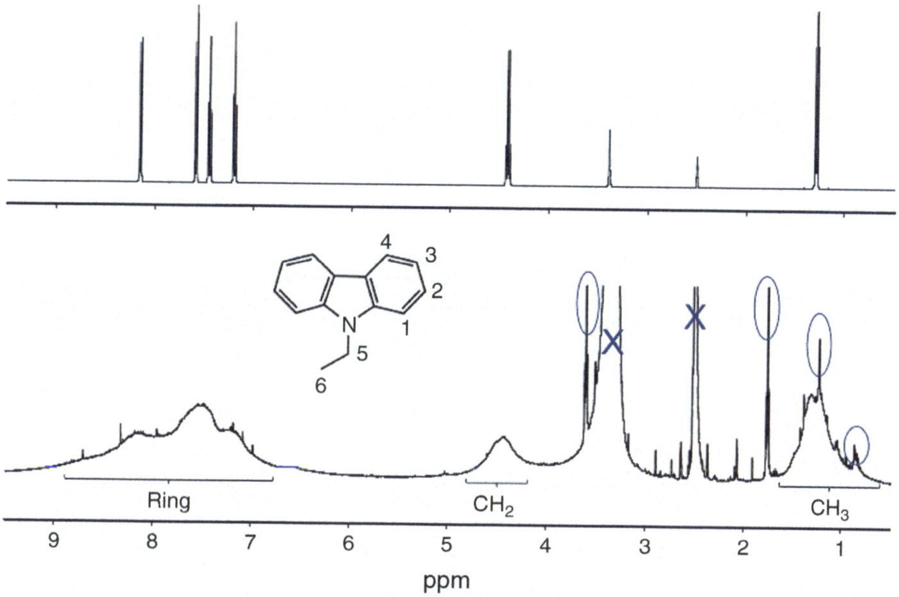

Fig. 4.29 The proton NMR spectrum of NEC-CDots in DMSO-d_6 (lower) is compared with that of NEC molecules (upper). In the former, the crossed peaks are associated with DMSO and water in the NMR solvent, and the circled sharp peaks are due to residual THF and hexanes from the cleaning of the dot sample (the peak integrations are relatively not so significant). The integrations for **CH_3**:**CH_2**:ring-**H** are 3:1.2:6.1. (Originally Figure 4 in Ren et al. (2019). Reproduced with permission)

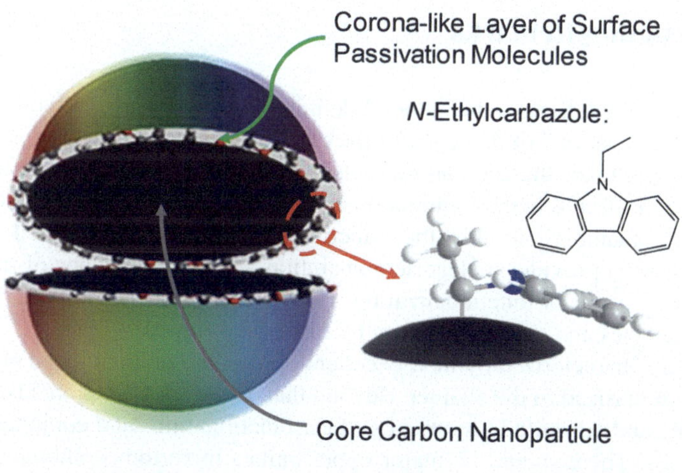

Fig. 4.30 A cartoon illustration on the structure of CDots in general, with more specifically the carbon nanoparticle surface binding and passivation by the carbazole species. (Originally Figure 1 in Ren et al. (2019). Reproduced with permission)

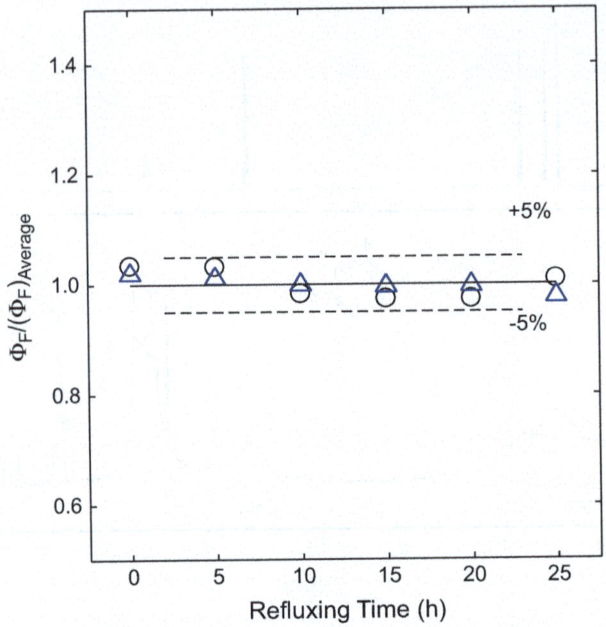

Fig. 4.31 The variations (deviations from the average represented by the solid line) in the observed fluorescence quantum yields of NEC-CDots at 400 nm (○) and 440 nm (Δ) excitations with respect to the dot sample being treated by refluxing in toluene for up to 25 h. The dashed lines define the range of +5% and −5% deviations from the average. (Originally Figure 7 in Ren et al. (2019). Reproduced with permission)

6 Concluding Remarks

The two critical elements in the general definition of CDots are "carbon nanoparticles" and "passivation" of defects. For the classical synthesis based on pre-existing carbon nanoparticles, there are no major issues with the first. On the second among different passivation schemes, amidation chemistry has served the purpose of being simple and unambiguous for the functionalization, which coupled with de-functionalization provides a clear demonstration on the significance of passivation effects. Other modes of functionalization, especially the radical addition, offer some major advantages and should therefore be further developed.

CDots are characterized by their excellent optical and photoinduced redox properties, as emphasized in the chapter. Beyond that, however, CDots are also structurally flexible and versatile in terms of surface functionalities and conjugations with other species. There should be major opportunities in various combinations of all these characteristics.

The experimental results have been very consistent with the proposed/assumed mechanistic framework of CDots. However, the "physics" (quantum mechanistic elucidation) on why the passivated defects in the carbon nanomaterials would lead to the observed properties, including the details on structure–property correlations, remains an extremely important and apparently also extremely tough challenge. It might require new theoretical understandings in solid-state physics.

Chapter 5
Other Syntheses: Popular Methods and Issues

CDots represent a nanomaterials platform in which the unique properties of small carbon nanoparticles are exploited and dramatically enhanced, namely, nanoscale carbon (in the form of particles or domains) is at the core of CDots conceptually and literally, dominating the dot structure and composition. Such a dominance is obviously ensured in the classical synthesis with pre-processed and selected small carbon nanoparticles for surface functionalization–passivation, as detailed in Chap. 4.

In general, the dot syntheses in the literature may be divided roughly into two categories, one based on the use of pre-existing carbon nanomaterials as precursors, including especially the classical synthesis discussed in Chap. 4, and the other on the carbonization of carbon-rich/carbon-containing species, often by "one-pot" processing. In the former, there is no question on the required dominance of nanoscale carbon, but for the latter, unfortunately, many or most of the reported carbonization syntheses have yielded dot samples that are carbon-deficient or even largely carbonless (Liang et al. 2019). Particularly troubling have been some of the claims on the carbonization synthesis of "red/near-IR carbon quantum dots" and "graphene quantum dots."

1 Syntheses with Pre-existing Carbon

There are a few synthetic approaches in which pre-existing carbon materials are used as precursor for the carbon nanoparticles or nanoscale carbon domains in CDots, in addition to the classical synthesis discussed in previous chapters. Among more popular has been the electrochemical etching of carbon materials. In the early study by Canadian researchers (Zhou et al. 2007), multiple-walled carbon nanotubes were electrochemically treated to produce "blue luminescent carbon nanocrystals" (Fig. 5.1) with an average size of 2.8 ± 0.5 nm in diameter (according to TEM imaging results). The photoluminescence emission spectra were found to be

© Springer Nature Switzerland AG 2020

Y.-P. Sun, *Carbon Dots*, https://doi.org/10.1007/978-3-030-41184-8_5

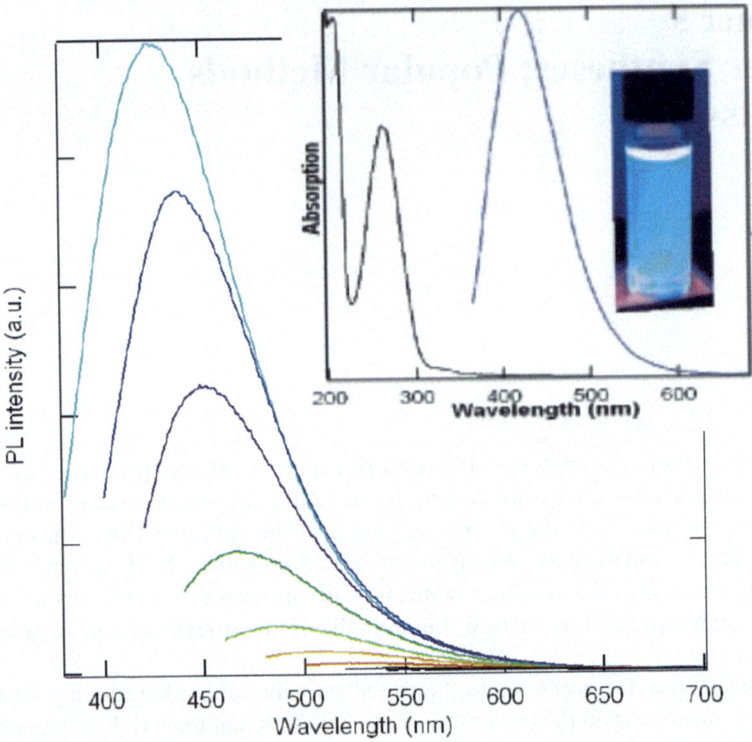

Fig. 5.1 Excitation-dependent photoluminescence spectra of carbon nanocrystals dispersed in deionized water. Inset: UV-vis absorption and photoluminescence spectra of carbon nanocrystals in aqueous solution. The photoluminescence spectrum was obtained under excitation at 365 nm, with the inset showing the solution illuminated by an UV lamp. (From Zhou et al. (2007). Reproduced with permission)

excitation wavelength dependent, though mostly in the UV region, suggesting that the "nanocrystals likely have variable emission sites and/or surface states on individual nanocrystals" (Zhou et al. 2007). The luminescence was rationalized as being defect-derived: "a widely accepted mechanism for luminescent nanocrystals is the radiative recombination of excitons from not only carbon nanocrystals of different sizes in the sample but also a distribution of different emissive trap sites" (Zhou et al. 2007). The observed quantum yield of 6.4% at 340 nm excitation was not so bad for the sample without any deliberate particle surface passivation, though the yields with excitations in the visible spectral region appeared to be much lower.

In similar electrochemical etching preparations, graphite rods were used as anode and cathode in ultrapure water or base solution under various current and/or potential conditions. The dot samples thus produced exhibited optical absorptions mostly comparable with those of carbon nanoparticles from carbon nanopowders or soot samples. Generally these dots were not further treated for surface passivation by organic species, so their observed photoluminescence emissions were weak, often with very low quantum yields for excitations at 400 nm and longer wavelengths.

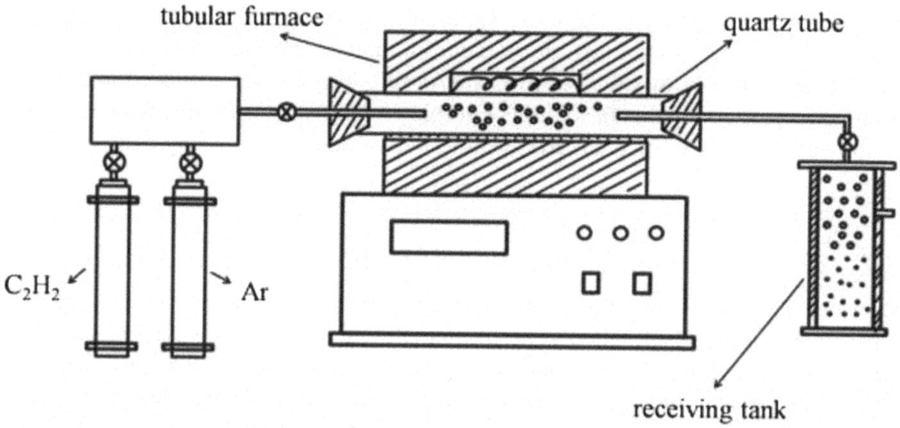

Fig. 5.2 Schematic of the CVD experimental apparatus for the preparation of carbon quantum dots. (From Yan et al. (2016) with permission)

The pyrolysis of small organic molecules in a CVD apparatus was used for the preparation of "carbon quantum dots" (Yan et al. 2016). In the study, acetylene was carbonized in quartz tube at 1000 °C under argon atmosphere (Fig. 5.2). The as-produced sample was dispersed in DMF and then filtered to collect the smaller dots. Without any further surface passivation, these dots should be similar to dispersed small carbon nanoparticles, with only rather weak photoluminescence emissions.

For the approaches highlighted above, the resulting "dots" are generally more like the pre-processed and selected small carbon nanoparticles in the deliberate chemical functionalization synthesis than the classically defined CDots. In principle, these dot samples may be further functionalized–passivated in various schemes for photoluminescence and related excited state properties similar to those that are characteristic of CDots.

2 "One-Pot" Carbonization Syntheses

The thermal or hydrothermal one-pot processing of organic or other carbon-containing precursors has dominated the dot syntheses. The implied purpose and assumption have been such that the organic precursors could be sufficiently carbonized under the processing conditions. However, little effort has been made to verify the assumption by quantifying the amount of nanoscale carbon actually formed in the carbonization processing. Some proofs, if any, for the presence of carbon nanoparticles have been transmission electron microscopy (TEM) images, but the fact of matter is that the powerful electron beams of up to 300 kV in TEM machines are more capable of carbonizing organic materials in situ during the imaging experiments than the reported carbonization processing conditions. It has also been suggested that some of the reported TEM images should be assigned to organic crystals formed on the TEM grid. There are apparently growing concerns in the research community on the contamination or even dominance in many or most of the dot samples by complex mixtures of dyes and/or dye-like species produced in the carbonization processing that have little or nothing to do with classically defined CDots (Xiong et al. 2018; Khan et al. 2018; Hinterberger et al. 2019; Liang et al. 2019).

2.1 *Contamination/Domination of Red/Near-IR Dye Mixtures*

CDots were originally pursued as performance-wise competitive yet nontoxic alternatives to conventional semiconductor QDs for bioimaging and related applications. A deficiency for such a purpose has been that optical absorptions of the small carbon nanoparticles as cores in CDots are relatively weak in the red/near-IR due to the characteristics of the π-plasmons, with correspondingly weak fluorescence emissions in the longer wavelength spectral region that is more suitable for bioimaging due to tissue transparency. Thus, the reports on syntheses of "red/near-IR carbon quantum dots" by hydrothermal processing of some specific organic mixtures at temperatures less than 200 °C for an hour or a few hours really wowed the research community. It was almost too good to be true, and unfortunately it was not true. In fact, those "red/near-IR carbon quantum dots" samples were heavily contaminated or more likely dominated by dye-like species/mixtures produced from the specific precursor mixtures under the processing conditions, as confirmed by results from more recent studies (Liang et al. (2019) and other results to be published, some of which are highlighted here).

For the red/near-IR absorptive and emissive samples from the carbonization of citric acid (CA)–formamide (FA) mixtures (Fig. 5.3), as a representative example, the reported mild processing conditions of 160 °C for an hour or a few hours were inefficient for the carbonization purpose but instead produced dye-like species/mixtures responsible for the observed red/near-IR absorptions and fluorescence emissions. Some of the experimental results for such a conclusion are provided and discussed as follows:

(i) The thermal treatments of various CA–FA mixtures for an hour or a few hours did yield samples of absorptions and emissions in the longer wavelength spectral region, different from those of CDots from the classical synthesis (Fig. 5.4). However, the absorptions of these samples were highly sensitive to even small variations in the processing conditions (Fig. 5.4). The fluorescence spectra and quantum yields were dependent on excitation wavelengths, but in patterns different from those for other CDots (Figs. 5.5 and 5.6).

(ii) The longer wavelength absorptions and emissions were specific to the precursor mixtures, completely absent for the samples obtained from CA–*N*,*N*-dimethylformamide (DMF, Fig. 5.3) mixtures, but observed for those from CA–*N*-methylformamide (MFA, Fig. 5.3) mixtures, all under the same thermal processing conditions (160 °C for an hour). The results on the exclusive

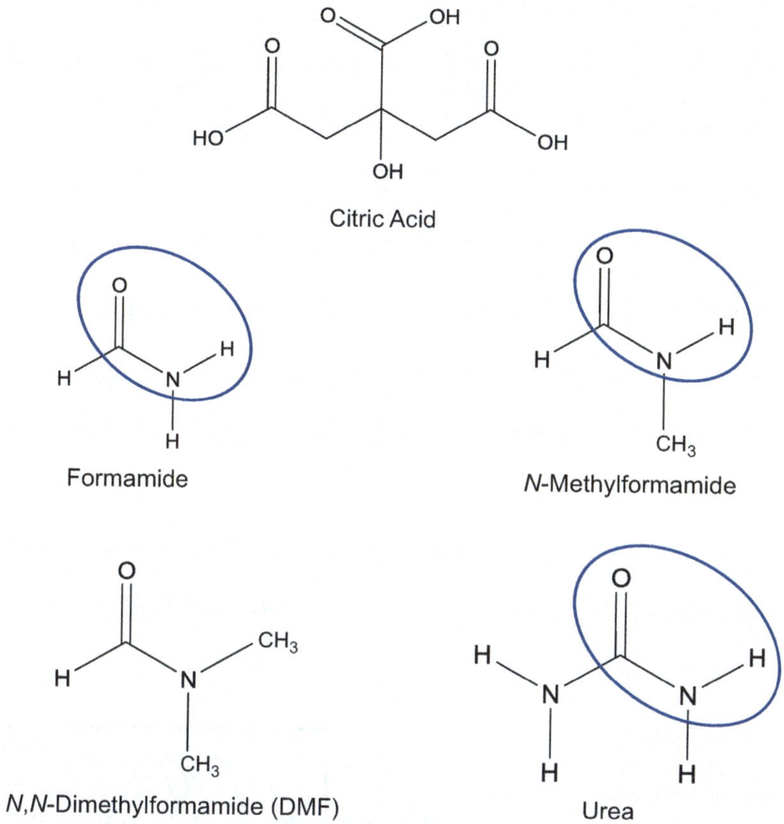

Fig. 5.3 Molecular structures of citric acid (CA), formamide (FA), *N*-methylformamide (MFA), *N,N*-dimethylformamide (DMF), and urea, with the –CONH structural unit highlighted

selectivity in precursors could only be explained by specific chemical reactions that require the –CONH chemical configuration (Fig. 5.3), because there could not be any exclusive selectivity for some minor changes in the chemical structure of the organic precursor under the blunt and indiscriminate thermal carbonization conditions. It is hardly surprising that the other reported precursor mixture for "red/near-IR carbon quantum dots" has been CA–urea (Fig. 5.3), because urea molecule also contains the –CONH structural unit (Fig. 5.3). As one would expect, studies of thermal reactions of CA with FA, MFA, urea, and/or other molecules containing the –CONH structural unit should be able to reveal the identities of the red/near-IR dyes that have dominated the absorptions and fluorescence emissions of the reported "red/near-IR carbon quantum dots."

(iii) The structures and compositions of the samples were found to be very complex. Upon dialyses of the samples with different pore sizes, there were always colored materials both inside and outside the dialysis tubing, which all

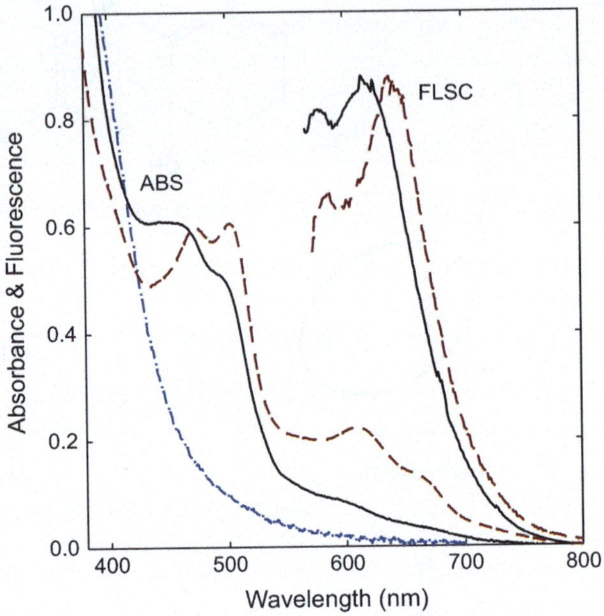

Fig. 5.4 Absorption (ABS) and fluorescence (FLSC, 560 nm excitation) spectra of the dot samples from the carbonization of citric acid–formamide mixtures at 160 °C for 1 h (experiment #1: —; and experiment #2: — —, with largely similar conditions in the two experiments) and at 300 °C for 2 h (-.-.-, ABS only)

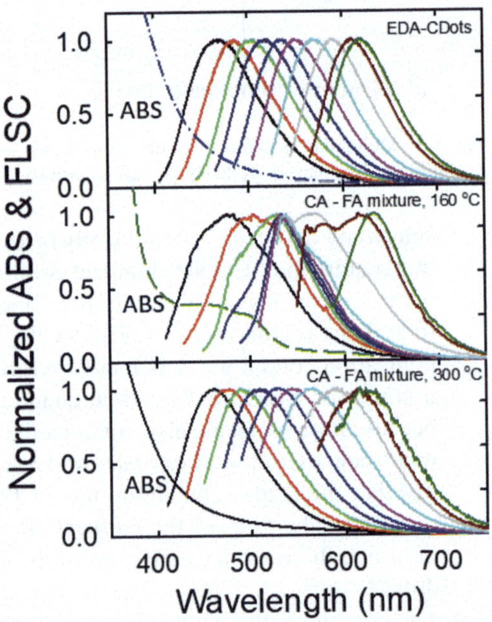

Fig. 5.5 Comparisons of absorption (ABS) spectra and fluorescence (FLSC) spectra at different excitation wavelengths (left to right from 400 nm in 20 nm increments) between EDA-CDots (upper) and the dot samples from the carbonization of citric acid–formamide mixtures at 160 °C for 1 h (middle) and 300 °C for 2 h (lower)

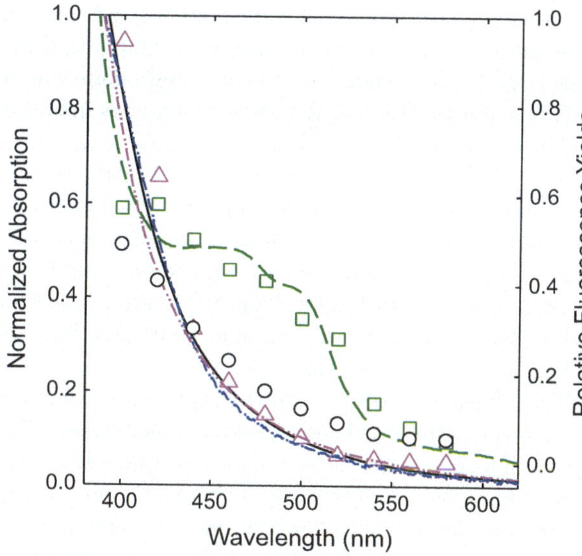

Fig. 5.6 Absorption spectra and fluorescence quantum yields at different excitation wavelengths for dot samples from the carbonization of citric acid–formamide mixtures at 160 °C for 1 h (— —, □) and 300 °C for 2 h (-..-..-, Δ). The results of EDA-CDots (——, O) and the absorption spectrum of aqueous dispersed carbon nanoparticles (-.-.-) are also shown for comparison

exhibited similar but quantitatively different absorptions and fluorescence emissions in the longer wavelength spectral region. There were apparently complex mixtures of free dyes in the samples, which were confirmed by results from NMR characterizations. The materials inside the dialysis tubing were likely "polymeric," namely, crosslinked organic species, some of which probably involved the small amount of nanoscale carbon domains formed under the processing conditions. One might visualize the sample morphology as being similar to "overcooked vegetables."

(iv) The seemingly extreme complexity with the samples, including the high sensitivity of the observed red/near-IR absorptions and fluorescence emissions to unintentional and unavoidable minor variations in processing conditions between repeating experiments (Fig. 5.4), fully dissipated when the samples were processed at higher temperatures for a longer period of time. For example, the same CA–FA mixture discussed above was processed at 300 °C for 2 h. The sample thus produced became rather simple in terms of its similar optical absorption and fluorescence emission properties to those of the classically synthesized EDA-CDots (Figs. 5.5 and 5.6) (Liang et al. 2019). Under those more robust processing conditions, the initially formed dye-like species/mixtures must have been further carbonized to yield higher contents of nanoscale carbon domains in the resulting samples, with the observed spectroscopic properties similar to those of the surface-functionalized/surface-passivated pre-existing small carbon nanoparticles. In fact, the application of the same more robust processing conditions to CA–DMF mixtures yielded samples of essentially the same spectroscopic properties.

The conclusion that the observed red/near-IR absorptions and fluorescence emissions must be due to dye-like species/mixtures in those samples from the specific organic precursors under mild processing conditions (more suitable for chemical reactions and insufficient for carbonization) is also consistent with the repeatedly observed dominance of π-plasmons in carbon nanostructures (including all classically synthesized CDots), such that the optical absorptions are little changed even with substantial modifications to the carbon nanostructures. It is simply impossible for "cooking some organic molecules for an hour or so" to create carbon nanostructures in which the dominance of π-plasmons could be altered so fundamentally to result in radically different electronic transitions. A cautionary note here is that carbon "quantum" dots are no magic, so if something seems too good or too easy to be true, it is simply not true!

There have been some interesting questions/comments on the dye contamination or dominance discussed above, such as "the red/near-IR emitting dyes could in principle be attached to the surface of the dots (namely carbon nanoparticles or nanoscale carbon domains), even by covalent bonds. In this case, someone might argue that it can be considered just as 'another type of functionalization' of the dot, hence a part of the dot in its own right, although in this case the absorption/emission mechanism would of course be entirely different, and probably more molecule-like than carbon-dot-like." As a simple answer, CDots attached covalently or noncovalently with organic dyes for strongly red/near-IR absorptive and fluorescent conjugates are known in the literature (Huang et al. 2013), so are CDots encapsulated/embedded with red/near-IR absorptive and fluorescent organic dyes (dubbed "host–guest CDots" (Sun et al. 2015)). These are not and cannot be considered as "red/near-IR carbon quantum dots" for obvious reasons. There have been extensive investigations in the literature on the modification of conventional semiconductor QDs with organic dyes. In such modified QDs, obviously the organic dyes are not and should not be considered as a part of the semiconductor QDs.

The so-called red/near-IR carbon quantum dots represent a more clear-cut example for poorly executed carbonization syntheses and misunderstandings of the produced samples. The carbonization processing of most other organic precursors and mixtures has not produced dot samples of red/near-IR absorptions and fluorescence emissions. For those dot samples, which have also been prepared under similarly mild carbonization conditions, it becomes more difficult to judge the sample quality. Nevertheless, those samples are still subject to the same problems due to the carbonization processing conditions being too mild to ensure the dominance of nanoscale carbon domains in the targeted dot structures.

2.2 *Insufficient Versus Sufficient Carbonization*

In the reported dot syntheses by the carbonization of organic precursors, there has generally been a major problem in terms of grossly underestimating the absolute need for robust processing conditions that are necessary for meaningful carbonization, let alone the required sufficient carbonization to ensure the dominant nanoscale carbon contents in the resulting dot samples. The commonly reported processing conditions of "cooking some precursor organic molecules or mixtures at less than 200 °C for a few hours" are far from being sufficient for the desired or assumed carbonization, as demonstrated, for example, in the carbonization of citric acid (CA)–oligomeric polyethylenimine (PEI) mixtures under several different processing conditions (Hou et al. 2017).

As reported (Hou et al. 2017), for the processing of a CA–PEI mixture under similar conditions to those that were used in a number of reported studies, 200 °C for 3 h under ambient pressure or hydrothermally in an autoclave at 150–200 °C, the resulting samples exhibited similar absorption and fluorescence spectral features to those reported in the literature (Fig. 5.7). The absorptions of these dot samples consisted of a broad band centered at around 360 nm, with little contributions in the visible spectral region, suggesting only minor presence of nanoscale carbon domains. On such an assessment, small carbon nanoparticles including nanoscale carbon domains are known to be broadly absorptive in the visible spectrum. The lack of substantial nanoscale carbon contents in the dot samples could be attributed to the too mild processing conditions significantly under-carbonizing the precursor CA–PEI mixtures. This was confirmed in subsequent experiments with the use of more vigorous processing conditions for the desired sufficient carbonization, including significantly higher temperatures and longer treatment times (Hou et al. 2017).

As a relatively simple modification to the popular carbonization protocol discussed above, a similar CA–PEI mixture in water was processed in a stainless steel

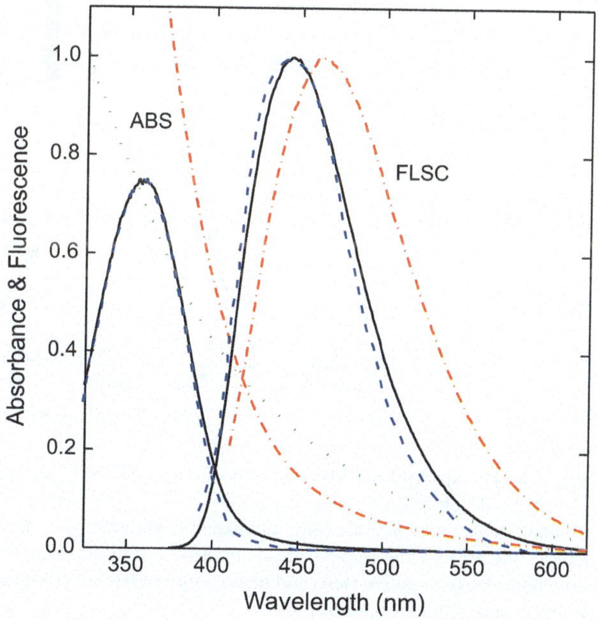

Fig. 5.7 A comparison of absorption (ABS) and fluorescence (FLSC) spectra of the dot sample obtained by following the protocol in the literature (200 °C for 3 h, ——) with those reported in the literature (- - - -). Also for comparison are the spectra of the PEI-CDots (-.-.-.-) and the absorption of aqueous dispersed small carbon nanoparticles (·······). (Originally Figure 2 in Hou et al. (2017). Reproduced with permission)

tube reactor at temperatures up to 400 °C for up to 12 h. It was found that the hydrothermal processing conditions around 330 °C for 6 h were most favorable, because at even higher temperatures and/or with longer hydrothermal treatment times, there was significant formation of insoluble precipitates. The dot samples obtained with the more favorable processing conditions (330 °C for 6 h) exhibited significant absorptions and bright fluorescence emissions in the visible spectral region (Fig. 5.8) (Hou et al. 2017), in good agreement with those of the CDots prepared by the classical synthesis with pre-processed and selected small carbon nanoparticles (Ge et al. 2018).

The results discussed above suggest that the samples prepared from carbonization syntheses under appropriate processing conditions could be comparable with those derived from pre-existing carbon nanoparticles, at least in terms of their observed optical spectroscopic properties. Here the appropriate conditions must be aggressive enough to ensure sufficient carbonization of the organic precursors for the dominance of nanoscale carbon domains in the resulting dot samples. Many researchers have apparently underappreciated the fact that it actually takes a lot to carbonize organic matters, as discussed in Chap. 2 on the difficulties encountered in the thermal carbonization of a small piece of beef patty under pretty harsh treatment conditions. However, there may also be technical issues with more aggressive

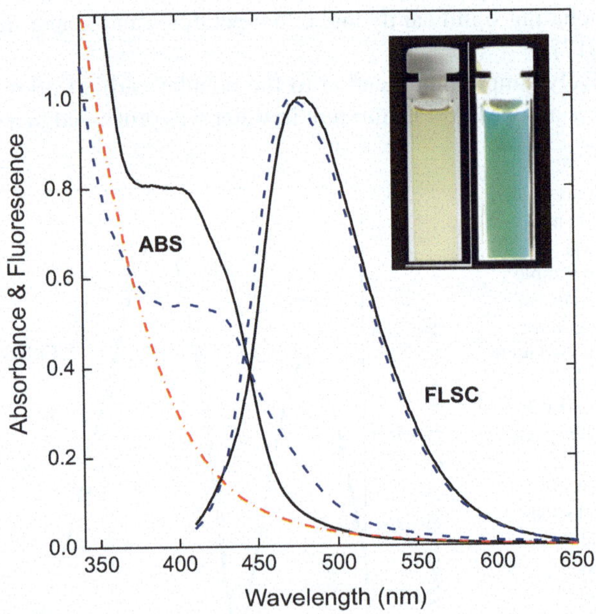

Fig. 5.8 The absorption (ABS) and fluorescence (FLSC) spectra of the PEI/CA-CDots prepared by the modified hydrothermal processing (——) are compared with those of the EPA-CDots obtained from the deliberate functionalization synthesis (- - - -). Also for comparison is the absorption of the PEI-CDots (-.-.-.-). Inset: Photos of a dilute aqueous solution of the PEI/CA-CDots under ambient room light (left) and under sunlight (right). (Originally Figure 3 in Hou et al. (2017). Reproduced with permission)

processing conditions, including the insolubility found in some of the samples thus obtained, which may have led to the widespread use of mild processing conditions in reported carbonization syntheses. The apparent conflict between the need for sufficient carbonization and the insolubility of some sufficiently carbonized samples may represent an intrinsic limitation of the one-pot carbonization synthesis method.

2.3 More on Carbonization and Resulting Dot Samples

The one-pot carbonization syntheses in reported studies were generally under processing conditions insufficient for the required carbonization. Nevertheless, the reported samples may be divided roughly into three categories: (1) grossly under-carbonized, so must be "carbon-deficient or even carbon-less carbon quantum dots," such as those from cooking organic molecules or mixtures at less than 200 °C for an hour or a few hours; (2) some, not many, prepared at higher temperatures with longer processing times, probably closer to classically defined and synthesized CDots in at least optical absorption and fluorescence emission properties; and (3) those in the middle in terms of processing conditions, likely as complex mixtures of some small carbon domains/pieces with plenty of organic species that are attached to or tangled with the carbon domains/pieces (namely, the mixtures being "polymeric" in nature due to the equivalent of crosslinking), thus impossible to remove abundant organic species via dialysis or other separation methods.

The dot samples in all three categories may be visualized as nanoscale versions of vegetables overcooked to different degrees (Fig. 5.9). The nanoscale carbon domains produced in the carbonization are like "knots" crosslinking some of the organic species that have survived the processing conditions. The under-carbonized samples may simply contain less and smaller "carbon knots," corresponding to overall lower nanoscale carbon contents in the samples. Morphologically, the available samples from carbonization syntheses must be intrinsically porous, which are likely responsible for some of their observed properties. For example, in the use of such samples for sensing metal ions based on the quenching of fluorescence intensities, some of the reported Stern–Volmer quenching constants were astronomically large, which could only be explained by the extreme local concentration enhancement due to sponge-like porous dot structures socking up the metal ions.

It should be noted that even the poorly prepared dot samples may exhibit some properties and/or can be used in some applications that are pursued for well-defined CDots, because organic dyes, including the dye-like species/mixtures in those samples produced during carbonization processing, may be able to perform many of the similar functions. For optical and related properties, CDots and QDs in general are in fact developed as more advantageous and superior alternatives to traditional dyes, so that it is hardly surprising for them and organic dyes to share some of the same behaviors and uses.

Fig. 5.9 Upper: Cartoon illustrations on likely structures of dot samples from the carbonization of organic precursors under different processing conditions (milder for the left and more vigorous for the right). These samples may be viewed as "composite-like," namely, the survived organic species crosslinked by nanoscale carbon domains. Lower: Potato thin slices overcooked in a stir-fry pan for the left and on an aluminum foil in an air fryer for the right

The emphasis on simple "one-pot" may have contributed significantly to the under-carbonization problems. Technically, it is very difficult to induce sufficient carbonization and to still have the resulting dot samples maintain the desired solubility characteristics. Except for some special cases based on the use of some specific precursors, the development and implementation of multistep processing protocols are required for more successful carbonization syntheses of CDots.

Even for dot samples synthesized by carbonization of organic precursors under appropriate processing conditions, it is unlikely for the dot structures in such samples (Fig. 5.9) to match those of CDots made from pre-existing solid carbon nanoparticles (Fig. 1.1). The difference in the nanostructures may be visualized in terms of a macroscopic analogy, such that the latter is more like "a hazelnut coated with chocolate" versus the former more like "chopped hazelnuts mixed with chocolate." On comparisons of their optical absorption and fluorescence emission properties, major agreements between the two kinds of dot samples have been found and reported (Ge et al. 2018). One may even argue that the former might be more advantageous in displaying brighter fluorescence emissions due to more passivated nanoscale carbon surfaces associated with the collection of more but size-wise smaller nanoscale carbon domains in the dot structure (Hou et al. 2017). Beyond the optical properties, however, the consequences of the nanoscopic structural differences to their other uses, such as in optoelectronic devices and for various treatments based on photodynamic effects, have yet to be investigated and understood in a systematic and comprehensive fashion.

The obvious dye contamination or dominance in the red/near-IR absorptive and emissive samples from the under-carbonization of some specific precursor mixtures represents one of the more extreme examples for the serious consequences of carbonization syntheses with inadequate processing conditions, as discussed above. Similarly extreme have been the various claims on the formation of crystalline graphitic nanoparticle structures (Fig. 2.19), or so-called graphene quantum dots, from the carbonization of similar organic precursors under equally inadequate processing conditions.

3 Carbon Dots or "Graphene Quantum Dots"

It would have to take a miracle to convert organic species into the kind of strange but nonetheless highly ordered graphitic nanoparticle structures (Fig. 2.19) with carbonization processing at 200 °C or less for an hour or a few hours. Yes, there have been transmission electron microscopy (TEM) images showing graphitic fringes, but the powerful electron beams in TEM machines are considerably more capable of creating graphitic structures in situ, as demonstrated in various reported studies of carbon nanomaterials. However, beyond those erroneous claims on the outcomes of the specific carbonization syntheses, there is the need for an understanding of the differences and relationships between CDots and "graphene quantum dots" (the quotation marks used purposely for the fact that the claims on dot samples obtained from carbonization syntheses have little or nothing to do with the originally defined graphene quantum dots (Ponomarenko et al. 2008; Westervelt 2008; Hewageegana and Apalkov 2017)). Here, some experimental evidence and facts relevant to such an understanding are presented to allow readers to make up their own minds and reach their own conclusions.

3.1 Originally Defined Graphene Quantum Dots

Simply, a small piece of single-layer graphene, in which the motion of electron is restricted (or confined as in conventional semiconductor QDs), was considered as a graphene quantum dot (see the review (Hewageegana and Apalkov 2017)). Such a dot configuration may be achieved by physical isolation in the "type I" or electronic confinement in the "type II" (Fig. 5.10) (Hewageegana and Apalkov 2017).

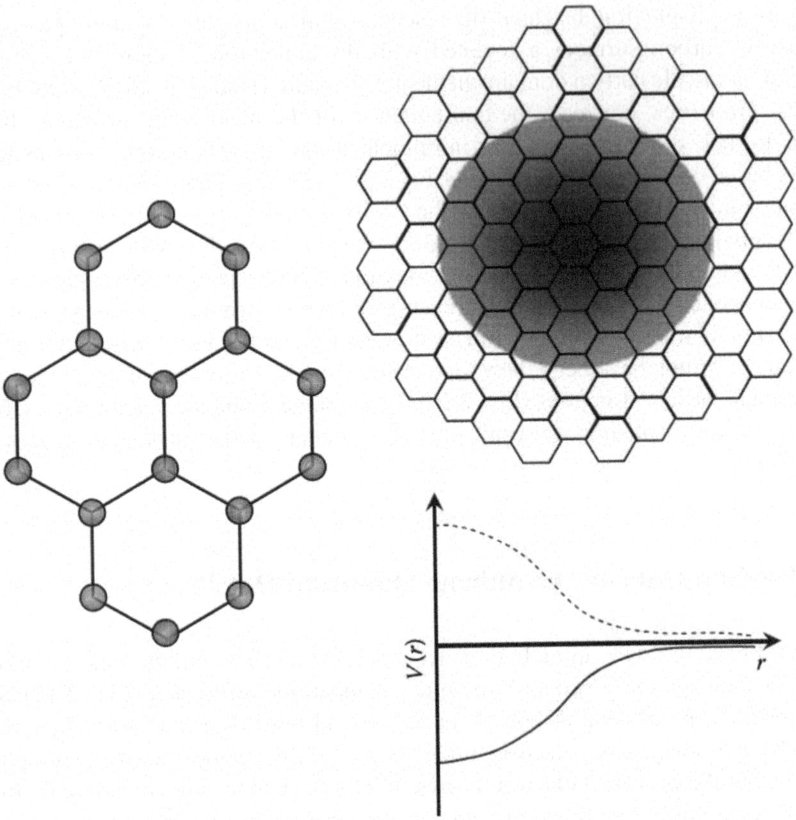

Fig. 5.10 Left: A type I quantum dot in graphene, consisting of a finite number of carbon atoms. Right: A type II quantum dot in graphene (upper) realized through a confinement potential, which (lower) can be of electron type (solid line) or hole type (dashed line). (From Hewageegana and Apalkov (2017). Reproduced with permission)

The type I graphene quantum dots were explored experimentally. As reported in (Ponomarenko et al. 2008), a polymer mask made with high-resolution electron-beam lithography was used to protect the selected graphene areas during the oxygen plasma etching to carve graphene into a desired geometry. Graphene quantum dots of 15–250 nm were obtained for their being evaluated as transistors. The results showed that the larger dots (>100 nm) behaved as conventional single-electron transistors, exhibiting periodic Coulomb blockade peaks, but for the smaller dots (<100 nm), the peaks became strongly nonperiodic, suggesting major quantum confinement effects. In a similar study (Westervelt 2008), the isolation of a graphene quantum dot containing only a few benzene rings was more clearly demonstrated (Fig. 5.11).

There has been no "bridge" that would convincingly connect these originally defined graphene quantum dots to the kind of size-dependent optical properties famously found in conventional semiconductor QDs, not for the visible spectral

Fig. 5.11 Graphene devices. (Left) Scanning electron micrograph of a graphene-based single-electron transistor. The dark areas are gaps in the photoresist mask where graphene is removed by plasma etching. The small quantum dot in the center is connected to contact regions through narrow constrictions, and there are four side gates. (Right) Schematic showing a hypothetical graphene-based single-electron transistor based on a quantum dot that contains only a few benzene rings. (From Westervelt (2008). Reproduced with permission)

region at the least. Unfortunately, the concept and definition of the original graphene quantum dots have been misused in various unsubstantiated claims, some of which are highlighted as follows:

1. In the structure of the reduced single-layer graphene oxide, as claimed there might be "sp^2 islands" resembling those of the type I graphene quantum dots, but the observed colorful fluorescence emissions in the visible spectrum could not be bandgap in origin, but more likely due to edge and structural defects in the reduced graphene oxide.
2. The so-called top-down preparation of "graphene quantum dots" from selected carbon precursors would yield carbonaceous particulates of some graphitic fragments in the structure, with their observed optical absorption and fluorescence emission properties mechanistically similar to those of carbon nanoparticles or poorly passivated CDots. Those from "dirty" carbon precursors like coal would also be subject to significant spectral effects due to colored contaminants.
3. In carbonization syntheses from organic precursors, the overwhelming majority of the claimed "graphene quantum dots" were prepared under processing conditions grossly insufficient for the creation of substantial nanoscale carbon domains, let alone any nicely defined single-layer graphene pieces. Even with considerably more vigorous carbonization for the formation of some small double- or multiple-layer graphene nanosheets, the claims for the graphene quantum dots-like entities in these species seem ridiculous in terms of photophysics 101. It is well established that in stacked π electron structures (like in stacked graphene sheets), the photoexcited states of one structure would be quenched highly efficiently (near-neighbor static) to diminish any fluorescence emissions. The observed colorful emissions in such nanomaterials must be defect-based/defect-derived, as discussed next.

3.2 Functionalized Graphene Nanosheets and Optical Properties

Graphite can be expanded and exfoliated into thin pieces of various aspect ratios. The exfoliation processing has often involved the use of oxidative acids, which could oxidize the carbons associated with structural and edge defects to form oxygen-containing moieties including carboxylic acids (Fig. 5.12) (Veca et al. 2009). As a result, some relatively smaller pieces such as around 5 nm in thickness and less than 500 nm in lateral dimension may be somewhat dispersible in polar solvents with rigorous sonication, though the resulting suspensions are generally unstable against gravity, with substantial precipitation of the few- or multi-layer graphene nanosheets mostly within only minutes. These nanosheets are hardly pristine due to the oxygen-containing moieties in the nanosheet structure and on the edges (Fig. 5.12), which are often considered as "functionalization" in some literature reports. Nevertheless, these samples generally exhibit major light scattering effects in optical absorption measurements, and only rather weak photoluminescence emissions both in suspensions and in polymeric matrices.

The lack of photoluminescence emissions has made Raman characterizations of the fewer-layer graphene nanosheets straightforward and more quantitative. However, the oxygen-containing moieties could be used as targets for chemical functionalization with polymeric species, such as poly(vinyl alcohol) (PVA) or

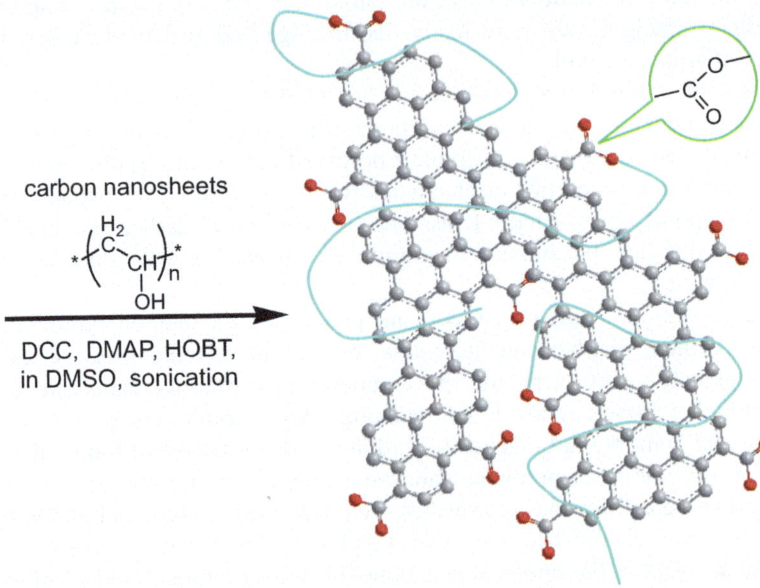

Fig. 5.12 The functionalization via esterification of the carboxylic acid moieties in oxidized carbon nanosheets (fewer-layer graphene nanosheets) by hydroxyl groups in PVA. (Originally Scheme 1 in Veca et al. (2009). Reproduced with permission)

other polymers bearing hydroxyl or primary/secondary amine groups. Upon the attachment of PVA polymer via the carbodiimide-activated esterification reaction, for example (Fig. 5.12), the functionalized graphene nanosheets could become significantly photoluminescent, to the degree of interfering with or even prohibiting Raman measurements (Fig. 5.13) and at the same time readily enabling the confocal fluorescence microscopy characterization (Fig. 5.14) (Veca et al. 2009).

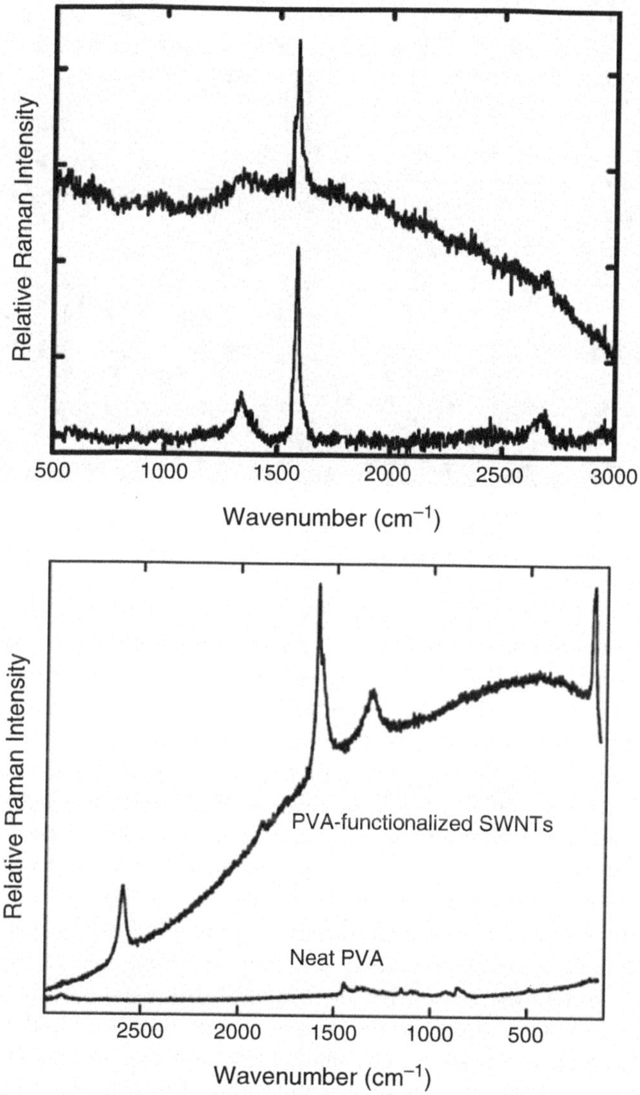

Fig. 5.13 Raman spectra of (UPPER) carbon nanosheets (fewer-layer graphene nanosheets) before (lower curve) and after PVA functionalization (upper curve) and (lower) PVA-functionalized SWNTs. (From Veca et al. (2009) and Lin et al. (2003). Reproduced with permission)

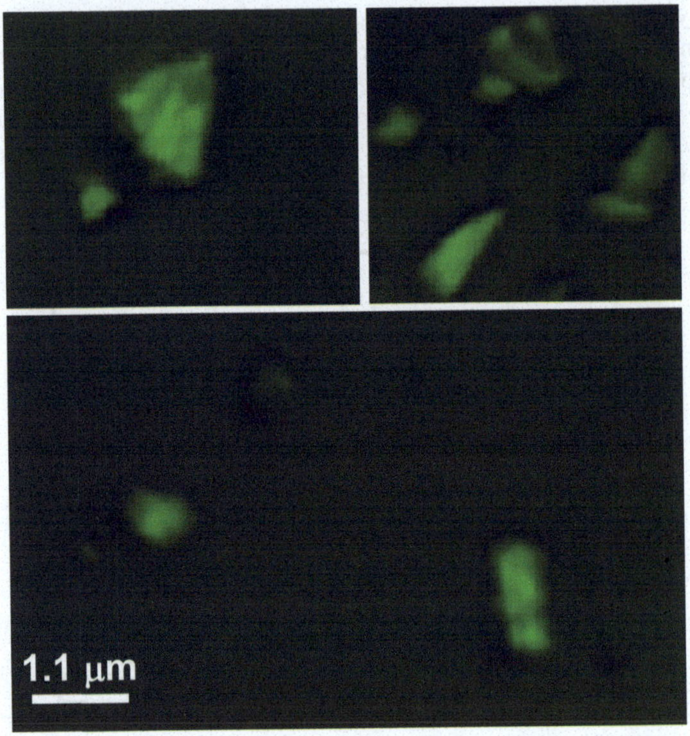

Fig. 5.14 Confocal microscopy (reflection mode) images of the PVA-functionalized carbon nanosheets (fewer-layer graphene nanosheets). (Originally Figure 3 in Veca et al. (2009). Reproduced with permission)

Similarly functionalized few-layer graphene nanosheets with aminopolymers such as PPEI-EI (Fig. 4.1) exhibited brighter photoluminescent emissions, comparable with those in the PPEI-EI-functionalized carbon nanotubes (Riggs et al. 2000). The observed emissions of polymer-functionalized graphene nanosheets could not be due to any bandgap transitions associated with the delocalized π system in the nanosheets, rather must be credited to the passivated structural and edge defects. It should again be noted here that according to the established understanding in photophysics, in small pieces of graphene nanosheets with two or more layers, the photoluminescence emissions of one layer (considered as a large π-conjugated molecule in some literature reports) would be quenched completely by the adjacent layer in an extremely efficient next-neighbor "dark quenching" process. This is also the reason for the required dispersion (or de-bundling) of single-walled carbon nanotubes to the individual nanotube level in the detection of bandgap fluorescence emissions. The same dispersion requirement is not applicable to defect-derived photoluminescence emissions of functionalized carbon nanotubes, as well as those of the polymer-functionalized few-layer graphene nanosheets (Veca et al. 2009).

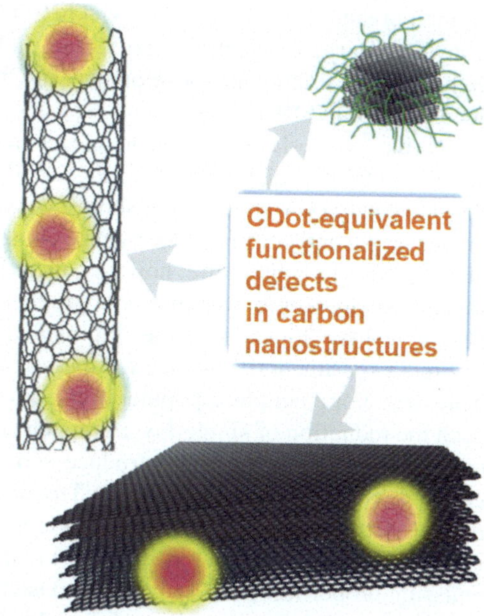

Fig. 5.15 A cartoon illustration on the CDot-like/equivalent functionalized defects in carbon nanomaterials, including the functionalized small graphitic pieces (upper-right)

Other than the different shapes and sizes, the carbon nanotubes and graphene nanosheets share characteristic structural features with small carbon nanoparticles in terms of their all containing similar carbon defects. The functionalization for effective passivation of the defects in all of these nanoscale carbon allotropes results in similarly bright and colorful fluorescence emissions, as well as similarly characteristic photoinduced redox behaviors. Thus, the functionalized-passivated defects in the carbon nanotubes and graphene nanosheets may conceptually be considered as an extended/special version of CDots (Fig. 5.15).

3.3 *"It Is All About Defects and Passivation"*

Figure 5.15 captures CDots-like defect-based/defect-derived photoluminescence emissions found experimentally in all carbon nanomaterials, including fewer-layer graphene nanosheets of different sizes, but obviously excluding fullerenes for the fact that they are nanometer-sized all-carbon molecules. The originally/classically defined graphene quantum dots (Ponomarenko et al. 2008; Westervelt 2008; Hewageegana and Apalkov 2017) cannot be applied to few-layer graphene nanosheets as some have suggested, at least in terms of their observed photoluminescence properties in the visible spectrum (which are no different from those of functionalized carbon nanotubes, Fig. 5.13, as both are derived from similar functionalized–passivated structural and edge defects). For those few-layer graphene nanosheets that are small and thus like nanoparticles, their surface passivation through organic functionalization yields CDots of more graphitic cores, which are still within the general definition of CDots. In fact, such dot structures were explored in the original finding and development of CDots and covered in the related patent document (Sun 2010).

On the single-layer graphene pieces representing the originally/classically defined graphene quantum dots (Figs. 5.10 and 5.11), their optical transitions should be in the UV spectrum or shorter wavelengths, with little practical relevance to the "bright and colorful fluorescence emissions" in the visible and extended into the near-IR. In the literature, however, there have been many claims on the synthesis and spectroscopic properties of "graphene quantum dots," "borrowing" (actually more like "hijacking") the original/classical concept and definition without any real experimental evidence for the true origins of the observed spectroscopic behaviors. For example, a popular preparation of the "graphene quantum dots" has been the reduction of single-layer graphene oxides, which would in principle (and probably in practice as well) produce some "sp^2 islands" resembling the type I graphene quantum dots (Fig. 5.10). Troubling is that there have been no experimental results (with little related efforts actually) to link or correlate these "sp^2 islands" and their electronic structures and transitions to the observed fluorescence emission properties. In fact, the mere existence of the "sp^2 islands" requires abundant defects to define the edges of the islands (Fig. 5.16) (Cao et al. 2013), whose potentially dominating contributions to the observed colorful fluorescence emissions must be considered and recognized. As already established for single-walled carbon nanotubes, their bandgap fluorescence emissions are relatively much weaker than the defect-based/defect-derived emissions, in addition to the fact that the former may be quenched or diminished by the defects. Therefore, many of the claims and assignments of "graphene quantum dots" being responsible for the observed spectroscopic properties in the literature may require re-examinations and reconsiderations.

As discussed in (Cao et al. 2013), "defects and their effects on electronic transitions are apparently at the center of both opportunities (for bright photoluminescence emissions) and complications relevant to the optical properties of graphene and related materials. Some in the carbon nanomaterials research communities have complained about confusions associated with how the observed photoluminescence is attributed or defined, such as how to differentiate fluorescence emissions found in specifically engineered graphene sheets and in graphene quantum dots. Adding to the complications and confusions is the fact that similar (at least phenomenologically and likely also mechanistically) photoluminescence emissions have been found in carbon nanotubes and small carbon nanoparticles (carbon dots). In those carbon nanomaterials, bandgap transitions and especially the critical role of structural defects and their passivation by various means have been explored extensively. Thus, there is a need for a systematic examination on photoluminescence emissions in these different yet related carbon nanomaterials toward a global view on the shared or distinctive mechanistic origins." The sentiment expressed at the time (Cao et al. 2013), as reflected in Fig. 5.16, was that "It is all about defects and their passivation." This is still valid today, namely, that the overwhelming majority of the

Fig. 5.16 Upper: Isolated sp^2 islands in a graphene sheet and a photo showing fluorescence in solution that was originally assigned to be bandgap in nature in the reported study but more likely to be defect-based/defect-derived and a multiple-layer graphene piece (left). Lower: a carbon nanoparticle with surface defects (left) and emission color variations in CDots. (From Cao et al. (2013). Reproduced with permission)

observed photoluminescence emissions in carbon nanomaterials, including various graphene structures and those claimed as "graphene quantum dots," must be due to defects in these nanomaterials, and the emissions can be substantially enhanced by the effective passivation of the defects.

4 Concluding Remarks

As increasingly concerned and recognized in the CDots research community, the proliferation of dot samples prepared by one-pot carbonization of organic precursors with the use of much too mild processing conditions (thus substantially contaminated with dye-like species/mixtures formed as thermal reaction products because of the insufficient carbonization) has become a major pollution, which damages the reputation of the research field and hinders its healthy advances. Some of these problems are likely due to the intrinsic limitations of the carbonization synthesis, such as the fundamental conflict between the need for sufficient carbonization and the poor solubility characteristics of the sufficiently carbonized products. Among the many cases presented and/or conclusions made in the chapter, the two most important ones are the following: (1) there are no "red/near-IR carbon quantum dots," as the spectral features displayed by the samples so claimed are due to red/near-IR dyes as products of thermal reactions (rather than carbonization); and (2) there is no evidence whatsoever for any correspondence or correlations between the observed bright and colorful fluorescence emissions and the classically defined graphene quantum dots or their equivalent nanostructures, and in fact the emissions and related photoexcited state properties are most likely associated with passivated defects in the graphene nanosheets.

Chapter 6
Representative Applications of Carbon Dots

CDots have been explored for a variety of potential applications. While the ideas behind the reported studies are mostly valid and interesting, the dot samples used in many of the studies may require some careful re-examinations on possibly major or even dominating effects of dye-like species/mixtures produced in carbonization syntheses of the samples under processing conditions that are insufficient for the required substantial nanoscale carbon domains (equivalent to the small carbon nanoparticles) in the generally defined CDots. Therefore, for the stated objective of not to have this book become another comprehensive review, only some selected uses of the CDots mostly derived from pre-existing carbon nanoparticles are highlighted in this chapter as representative examples to demonstrate the value and potential of CDots-based/derived technologies. More comprehensive accounts of all reported applications are available in many review articles, which are listed in the bibliography.

1 Bioimaging In Vitro and In Vivo

CDots were originally found and developed for their bright and colorful fluorescence emissions, and they were considered from the very beginning as potentially benign and nontoxic alternatives to the established conventional semiconductor QDs. In this regard, CDots have done very well in terms of both the nontoxic affirmation based on extensive toxicity evaluations and the high performance in bioimaging.

© Springer Nature Switzerland AG 2020
Y.-P. Sun, *Carbon Dots*, https://doi.org/10.1007/978-3-030-41184-8_6

1.1 Nontoxic

According to all available experimental results, CDots representing a new carbon-derived nanomaterials platform are intrinsically nontoxic or more precisely no more toxic than the organic species used for the surface passivation function. For example, the PEG_{1500N}-CDots sample used in many property and bioimaging studies was evaluated in cytotoxicity assays using human breast cancer MCF-7 and human colorectal adenocarcinoma HT-29 cells and also in mice, with the results confirming the nontoxic nature in vitro and in vivo (Yang et al. 2009a, c).

Cytotoxicity evaluations of other CDots derived from pre-existing carbon nanoparticles with a series of cell lines, including stem cells, also supported the conclusion that CDots are nontoxic (Wang et al. 2011; Liu et al. 2019).

In the experiments for toxicity evaluation in mice, the PEG_{1500N}-CDots sample with ^{13}C-enriched core carbon nanoparticles was used for the quantification by isotope-ratio mass spectroscopy. As reported (Yang et al. 2009c), the mice in two groups were exposed to two different dosages of CDots, and the third group exposed to 0.9% NaCl aqueous solution as control. All of the mice survived during the 4 weeks of the experiment, without any abnormal behaviors and clinical symptoms. "Established serum biochemistry assays were used to evaluate more quantitatively the influence of CDots on the exposed mice, including especially those for potential heptic injury and kidney functions. As shown in Fig. 6.1, the two important hepatic indicators, alanine amino transferase (ALT) and aspartate amino transferase (AST), were at similar levels for the mice exposed to different dosages of CDots and the control. The three indicators for kidney functions, uric acid (UA), blood urea nitrogen (BUN), and creatinine (Cr), were also unchanged among the three groups of mice (Fig. 6.1). These results suggest no toxicity of CDots in mice at exposure levels beyond those commonly used in optical imaging *in vivo* and for relatively long exposure times (up to 28 days). In any case, the *in vivo* toxicity of CDots, if any, is considerably lower than that of CdSe-based QDs, for which detectable toxic effects were reported to be at exposure down to the pico-mole level" (Yang et al. 2009c). According to histopathological analyses of the organs harvested from the mice post-exposure to CDots (Fig. 6.2), structures of the organs were normal, without steatosis, necrosis, or hydropic degeneration in the hepatic sections. "Typical splenic unit and lymphocyte were observed clearly in the spleen sections (Fig. 6.2). Similarly, in the sections of kidneys, the glomerulus structure could be distinguished easily. No protein liquid or necrosis was found in all groups" (Yang et al. 2009c).

As with carbon nanomaterials in general (Yang et al. 2007, 2008, 2009b; Liu et al. 2014), the amounts of CDots in the liver and spleen were found to be higher than those in other organs, but still low in absolute population. For example, in specimens of dissected liver and spleen harvested 6 h post-exposure, only a few scattered spots were detected in the fluorescence imaging (Fig. 6.2), which might be attributed to trapped CDots. These and the results from isotope-ratio mass spectroscopy (MS) analyses suggested rather minor accumulation in liver and spleen. In the MS study (Yang et al. 2009c), the clearance time of the CDots from mice was

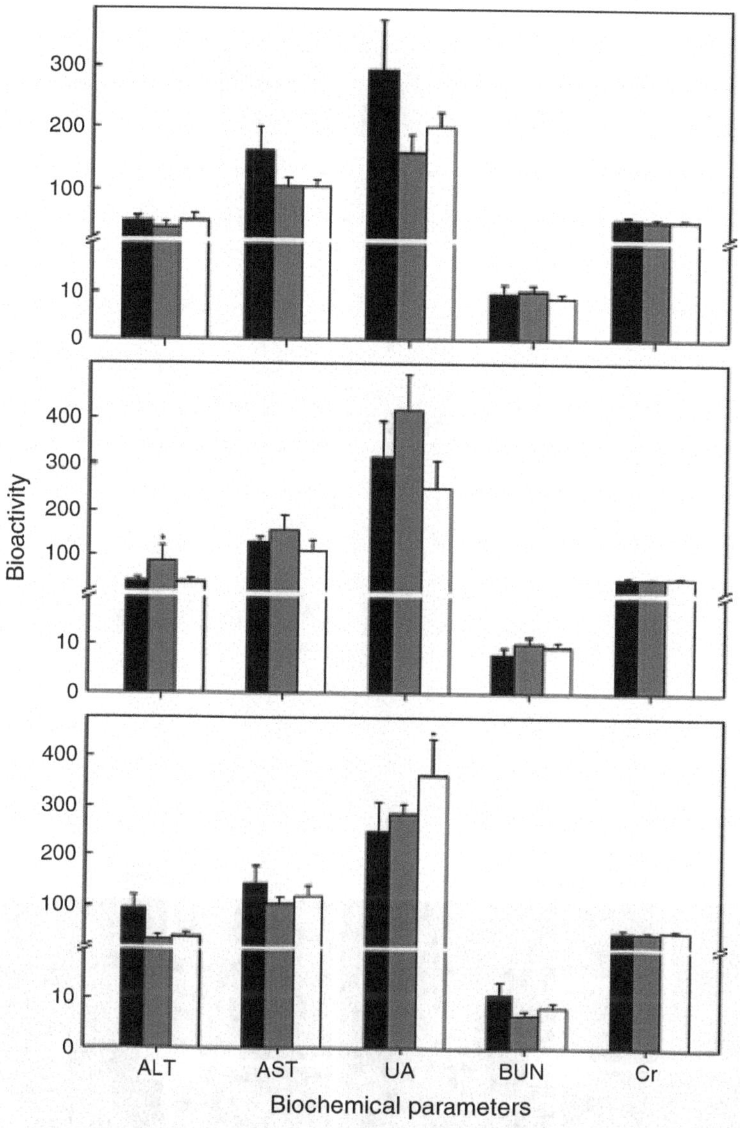

Fig. 6.1 Serum biochemistry results for mice intravenously exposed to PEG_{1500N}-CDots at carbon core-equivalent of 8 mg/kg (grey) and 40 mg/kg (white) and the control mice (black) at 1 day (top), 7 days (middle), and 28 days (bottom) post-exposure. Data presented as mean ± SD (n = 5). (Originally Figure 4 in Yang et al. (2009c). Reproduced with permission)

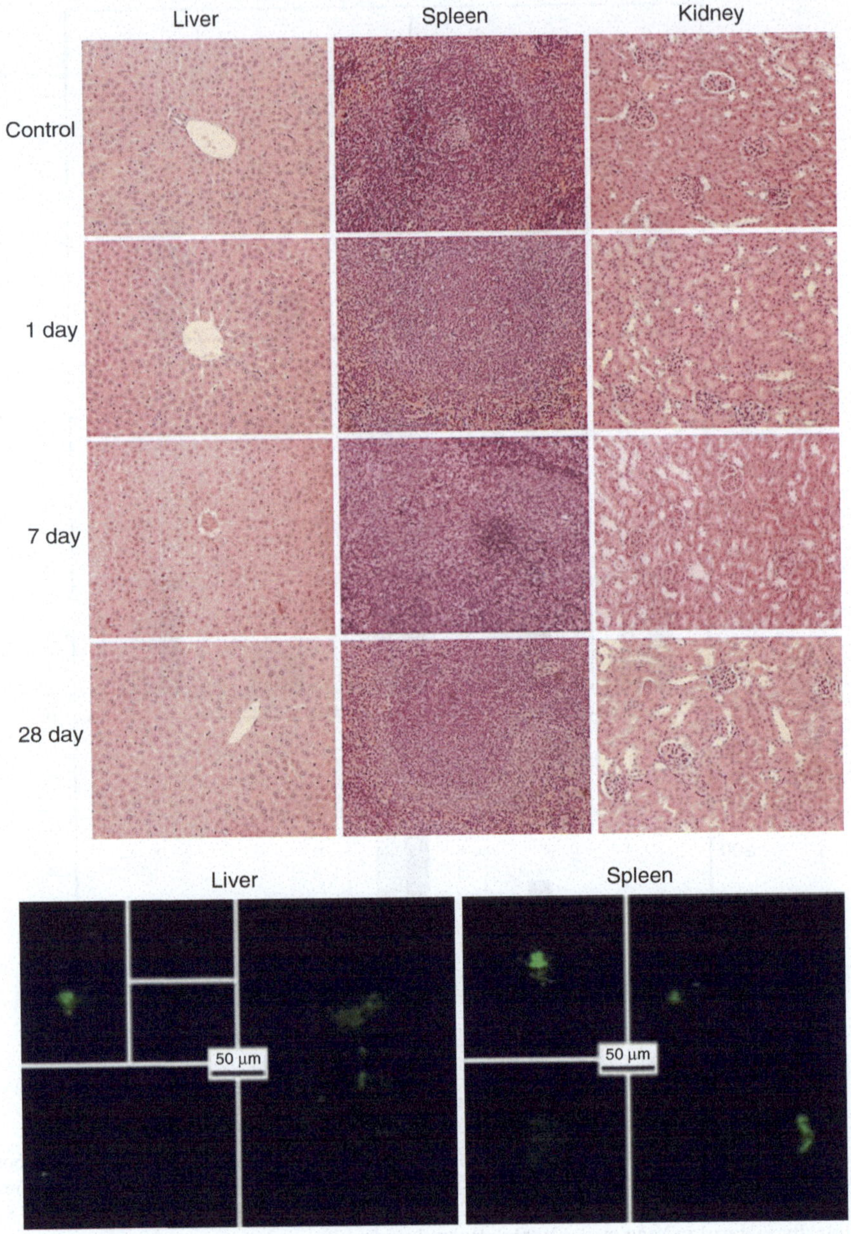

Fig. 6.2 Upper: Results from histopathological analyses of the liver, spleen, and kidneys. Lower: Fluorescence images (two-photon excitation at 800 nm) of sliced liver and spleen harvested from mice 6 h after intravenous exposure to PEG_{1500N}-CDots. (Originally Figure 5 in Yang et al. (2009c). Reproduced with permission)

estimated to be around 6 h, which was supported by the in vivo fluorescence imaging results from a separate investigation (Yang et al. 2009a).

1.2 Cell Imaging

In the original finding of CDots (Sun et al. 2006; Cao et al. 2007), their uses as regular (normal one-photon) and two-photon fluorescence probes for cell imaging were demonstrated. Since then, cell imaging has been a popular topic in the study of CDots. Highlighted here is a study of the well-established PEG_{1500N}-CDots for the regular and two-photon fluorescence labeling of cells (Liu et al. 2015). The human breast cancer MCF-7 and human colon adenocarcinoma HT-29 cells were selected for the labeling. For both cell lines, the fluorescence images acquired with 458 nm excitation (argon ion laser line) suggested significant uptake of the CDots by the cells, with the dots mostly residing in the cell membrane and cytoplasm and without any meaningful presence in the cell nucleus (Fig. 6.3). Qualitatively the brightness

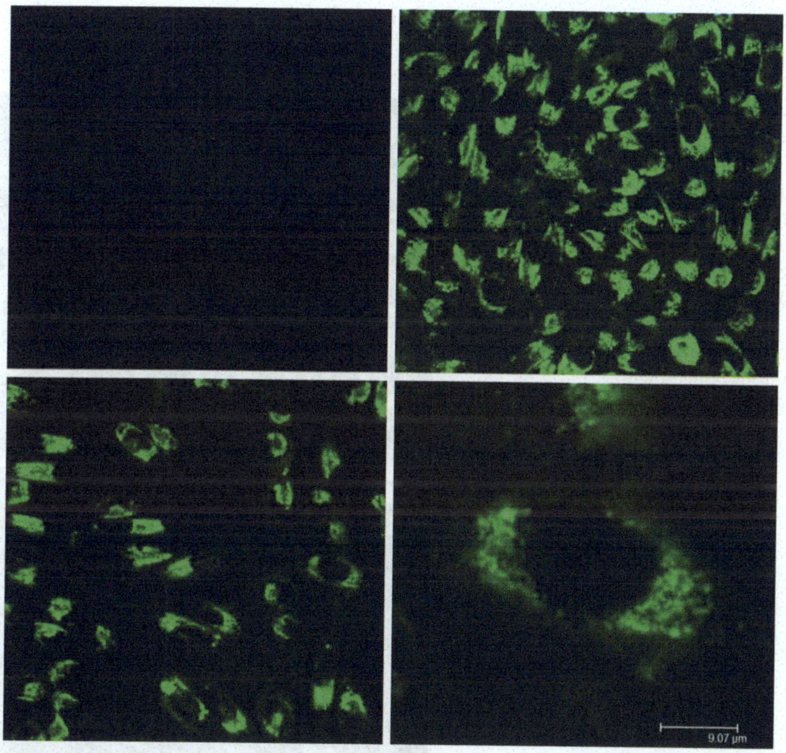

Fig. 6.3 Confocal fluorescence images (458 nm excitation, 470–590 nm emissions) for MCF-7 cells without (top left) and with the PEG_{1500N}-CDots (top right) and for HT-29 cells with the PEG_{1500N}-CDots at different resolutions (bottom left and right). (Originally Figure 6 in Liu et al. (2015). Reproduced with permission)

of the CDots in the fluorescence images of the cells was comparable with that in the images without cells, suggesting that the fluorescence properties of the CDots were not degraded in any significant fashion upon their being taken up by the cells.

The same cells labeled with PEG_{1500N}-CDots were imaged by using the fluorescence microscope with two-photon excitation (femtosecond pulsed laser at 800–900 nm) under otherwise the same experimental conditions. Green fluorescence emissions from the CDots in both cell lines could readily be detected, with the results again suggesting that the dots resided mostly in the cell membrane and cytoplasm (Fig. 6.4).

In an opinion article in *Nature Methods* by Baker (2010) on nanotechnology enabled probes for the emerging needs in cell imaging, CDots was featured as one of a few platforms for "smaller and more stable" imaging probes, where the small was defined as size-wise comparable to green fluorescent protein (GFP) (Baker 2010). Indeed, CDots could be made very small, such as EDA-CDots for ultra-compact fluorescence probes (Fig. 6.5) (LeCroy et al. 2014). The small probes are desired for minimally affecting the sub-cellular structure, in addition to being non-toxic, particularly suitable for fluorescence labeling of stem cells.

In the reported study (Liu et al. 2015), the SD rat mesenchymal stem cells (MSCs) within passage 8 were used in the cell labeling experiments with EDA-

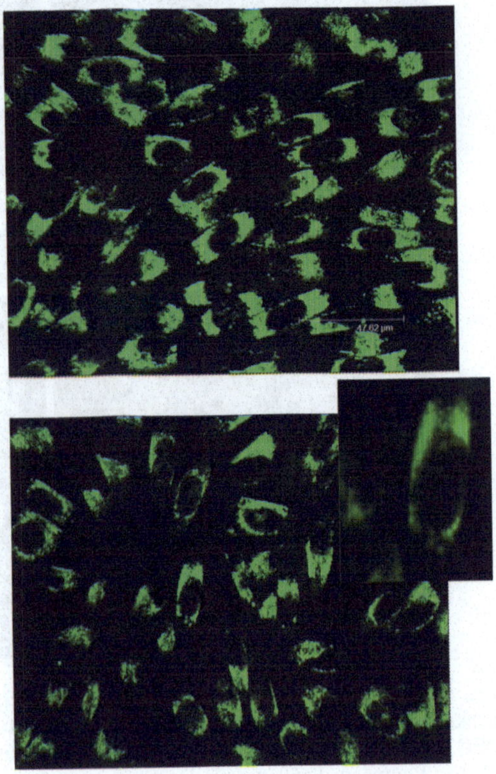

Fig. 6.4 Fluorescence images with two-photon excitation (800 nm excitation, 470-590 nm emissions) for MCF-7 (top) and HT-29 (bottom, and the inset) cells, both labeled with the PEG_{1500N}-CDots. (Originally Figure 7 in Liu et al. (2015). Reproduced with permission)

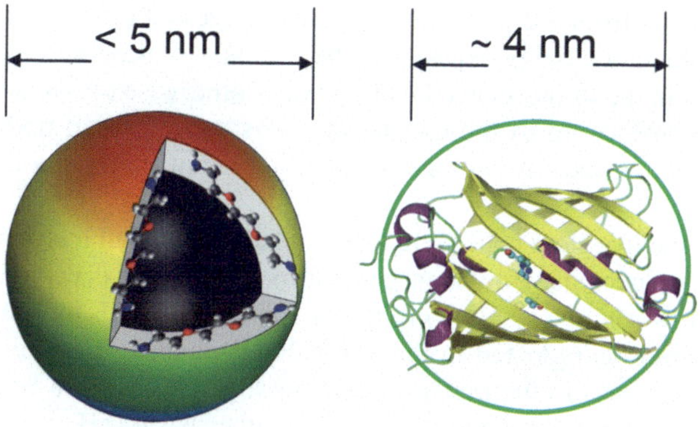

Fig. 6.5 A cartoon illustration on (left) an EDA-CDot, which is essentially a special "core-shell" nanostructure with small carbon nanoparticle as the core and a soft shell of tethered EDA molecules, and (right) green fluorescent protein with the size profile highlighted. (Originally Figure 1 in (LeCroy et al. 2014). Reproduced with permission)

Fig. 6.6 The merged (fluorescence + bright-field) images of the live stem cells labeled with EDA-CDots (upper and middle) and the control (without the dots, lower). (Originally Figure 8 in Liu et al. (2015). Reproduced with permission)

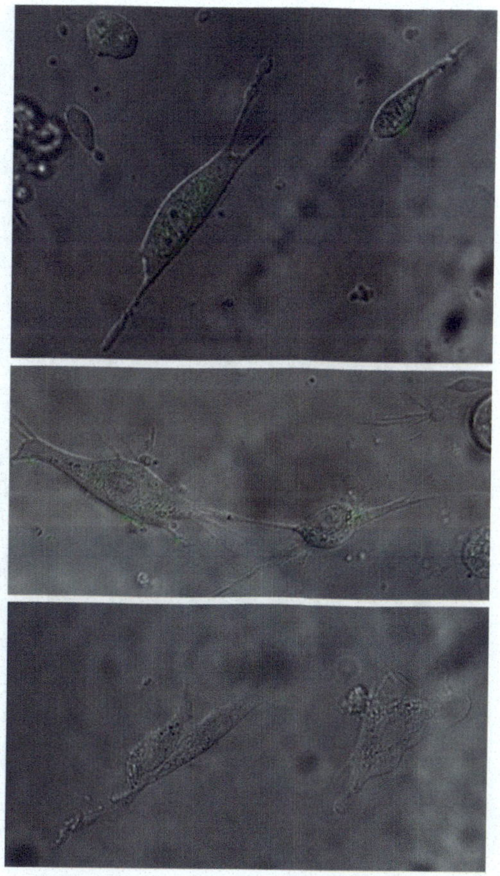

CDots. The results from the cell viability assay suggested that the stem cells were completely viable upon being exposed to the CDots at concentrations much higher than those used in the fluorescence labeling under otherwise the same experimental conditions. Post-exposure to the CDots, the cells were imaged under a confocal fluorescence microscope with 405 nm excitation to confirm the fluorescence labeling (Fig. 6.6).

The dots in the live stem cells were again mostly in the cell membrane and cytoplasm, not in the nucleus, similar in general to the cellular distribution of human cancer cells. However, the labeling efficiency for the stem cells by EDA-CDots was significantly lower, as reflected in the low cellular uptake of the dots and weaker fluorescence emissions in the images. "One possible cause might be related to the chemical structures of EDA-CDots, which are rather basic (pH ~13) due to a relatively large population of amino groups on the dot surface. Consequently, aqueous EDA-CDots sample had to be neutralized with an acid (aqueous HCl) before the stem cell labeling. The acid treatment protonated the amino groups to result in a similarly large population of cationic moieties on the dot surface, which were prob-

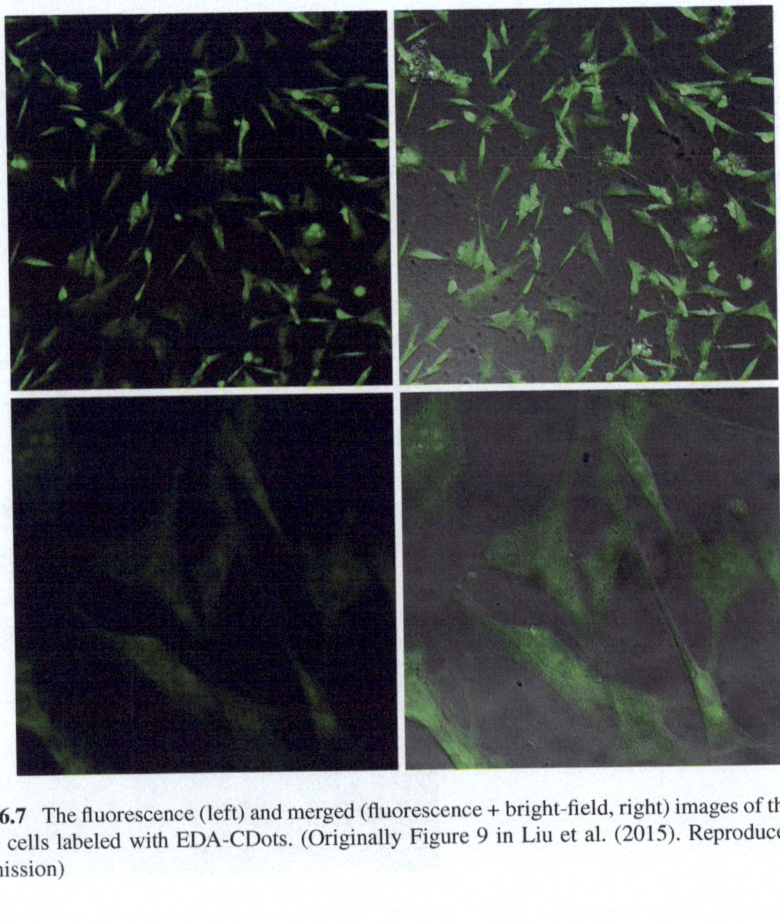

Fig. 6.7 The fluorescence (left) and merged (fluorescence + bright-field, right) images of the fixed stem cells labeled with EDA-CDots. (Originally Figure 9 in Liu et al. (2015). Reproduced with permission)

ably not favorable to the desired efficient uptake by the live cells. In a follow-up exploration on this issue, the MSCs were fixed by the treatment with 4% formaldehyde. The labeling of the fixed MSCs by the EDA-CDots was much more efficient, corresponding to much brighter fluorescence images and with the hyperchromatic nucleolus brighter than the other regions (Fig. 6.7)" (Liu et al. 2015).

The use of PEG_{1500N}-CDots with STimulated Emission Depletion (STED) microscopy for super-resolution fluorescence imaging of cells was demonstrated (Leménager et al. 2014). The dot sample with the broad emission spectrum tailing into the red allowed efficient STED at 592 nm. The comparison of the confocal image with the STED counterpart (Fig. 6.8) showed a strong improvement of the lateral resolution, breaking the diffraction limited resolution and improving the imaging quality. In a comparison of the measured average resolution values, the confocal mode was at 192 nm, a spot size close to the theoretical diffraction limit of confocal microscopy, while the STED mode was at 71 nm, with the best lateral resolution as low as 30 nm, far below the limit of the confocal resolution (Leménager et al. 2014).

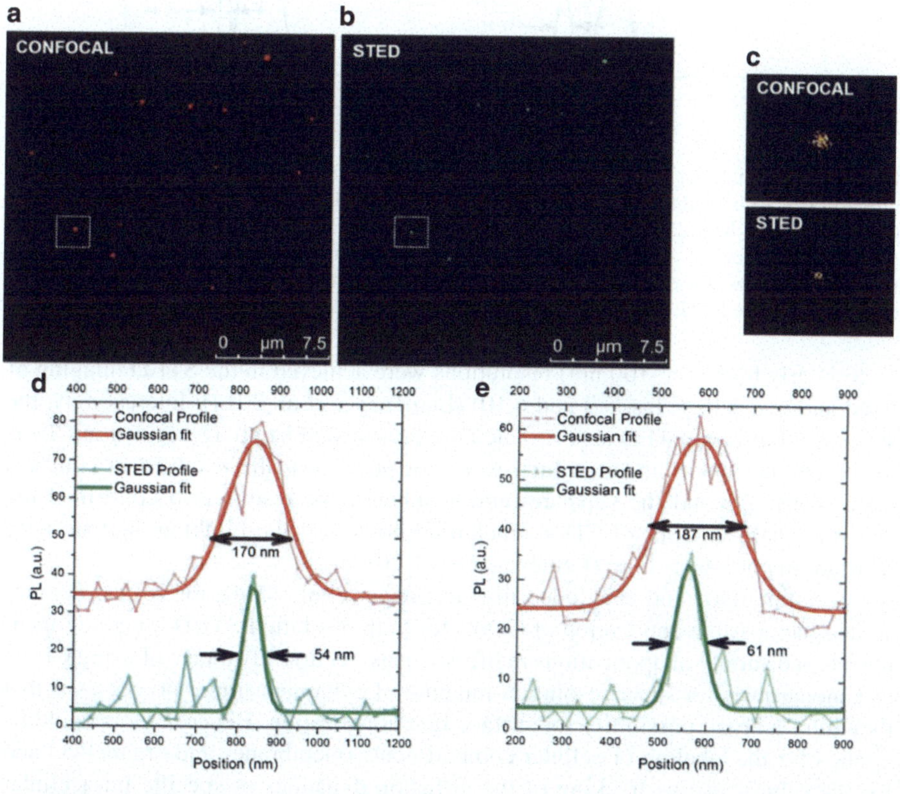

Fig. 6.8 Typical raw images of CDots observed with (**a**) confocal and (**b**) STED microscopy. (**c**) Zoom in view of the regions indicated in (**a**) and (**b**), respectively. (**d and e**) Graphs showing raw data (light color and symbols) and Gaussian fit (thicker line) of two representative intensity plots of fluorescent spots in confocal (red squares) and STED (green triangles) microscopy. (Originally Figure 2 in Leménager et al. (2014). Reproduced with permission)

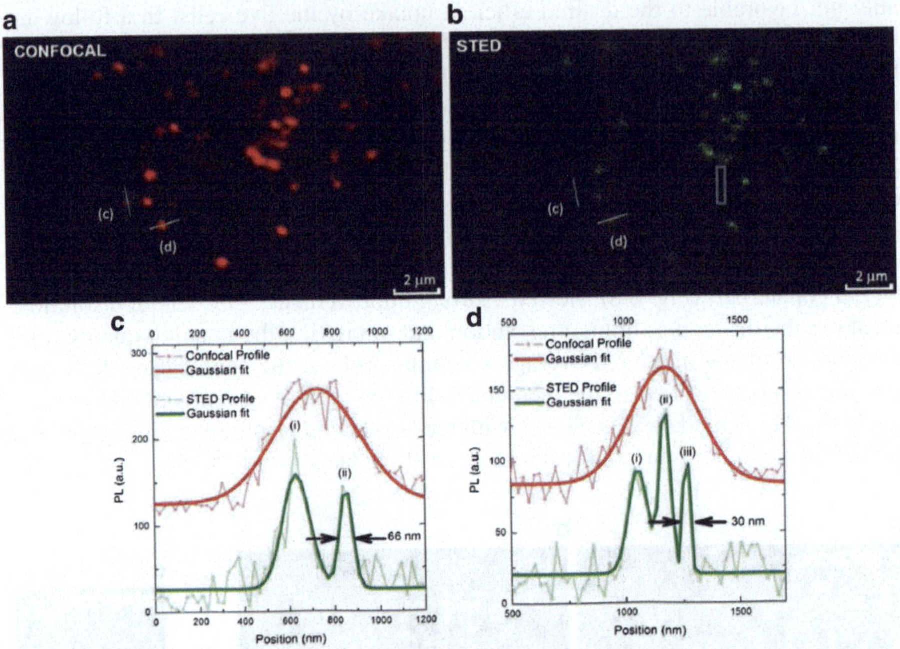

Fig. 6.9 Confocal (**a**) *versus* STED (**b**) imaging of CDots in fixed MCF-7 cells (CDots concentration: 170 nM, 48 h incubation). (**c** and **d**) Two representative line profiles extracted from signals of CDots in the cell. Raw data (light color with symbol) and Gaussian fits (thicker line) are shown in confocal (red color and square) and STED (green, triangle) mode. The FWHM values extracted from the Gaussian fit are (**c**) 140 nm (i), 66 nm (ii) and (**d**) 98 nm (i), 54 nm (ii) and 30 nm (iii). (Originally Figure 5 in Leménager et al. (2014). Reproduced with permission)

Similarly high (sub-100 nm) resolutions were achieved in the STED imaging of fixed and live cells (Figs. 6.9 and 6.10) (Leménager et al. 2014). Interestingly, the super-resolution of STED also revealed the presence of small aggregates of CDots in the cellular environment, which were not possible to be resolved in confocal microscopy. Beyond the super-resolution images, the results also confirmed the valuable characteristics of CDots, including nontoxic, non-blinking, and strongly resistant to photobleaching (Leménager et al. 2014).

As summarized and concluded in (Leménager et al. 2014), the results demonstrated the possible application of CDots for high resolution STED imaging, opening a broad variety of applications in life sciences. "A first advantage of using CDots as nanoparticles for super-resolution imaging of dynamic cellular processes is that they offer a broad possibility for surface functionalization. Hence, CDots could be applied for the labeling of cellular compartments (membranes and organelles) and for the sub-diffractive tracking of the diffusion dynamics of specific intracellular proteins over time. Another important advantage is that, unlike cadmium-based quantum dots, CDots are highly biocompatible, so they could be useful for the detailed visualization of cellular events over long periods of time or in *in vivo* experiments" (Leménager et al. 2014).

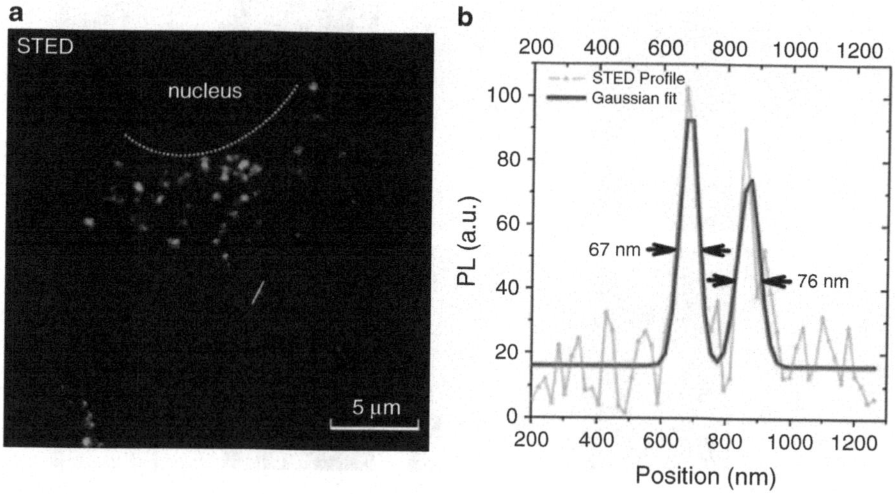

Fig. 6.10 Live cell imaging of CDots by STED microscopy. (**a**) STED imaging of CDots in MCF-7 cells incubated with CDots for 48 h. The white dotted line delineates the nuclear region. (**b**) Line profile (light green with triangle) and Gaussian fit (green) of STED CDots signals indicated with the white line in panel (**a**). A FWHM resolution of 67 and 76 nm was obtained. (Originally Figure 6 in Leménager et al. (2014). Reproduced with permission)

1.3 In Vivo Imaging

In the first in vivo imaging study of CDots (Yang et al. 2009a), aqueous solutions of PEG_{1500N}-CDots and C_{ZnS}-Dots were used for injection into mice in three different modes: subcutaneous, interdermal, and intravenous. For the subcutaneous injection, two excitation/emission wavelength pairs were used and compared (Fig. 6.11). The interdermal injection was to use the bright green fluorescence to track the migration through lymph vessels. The injected CDots did migrate along the arm (Fig. 6.12). As discussed (Yang et al. 2009a), "unlike semiconductor quantum dots such as CdSe/ZnS, which can migrate to axillary lymph nodes in minutes, the observed migration of the carbon dots was slower. This could be due to the small sizes of carbon dots (on the order of 4–5 nm) and/or the surface functionalization by the PEGs, whose protein resistance characteristics might reduce interactions of the carbon dots with lymph cells." The axillary lymph nodes were harvested and dissected at 24 h post-injection, in which the dot fluorescence could readily be detected (Fig. 6.12).

For the intravenous injection, the abdomen of mice was shaved for fluorescence detection of the dots trapped in organs during the whole-body circulation. However, only fluorescence emissions from the bladder area were observed (Fig. 6.13). At about 3 h post-injection, bright fluorescence in the urine became visible, suggesting that the intravenously injected CDots were excreted primarily via urine. The organs were harvested at 4 h post-intravenous injection for imaging analyses ex vivo. Only the dissected kidneys and liver exhibited meaningful fluorescence from the CDots,

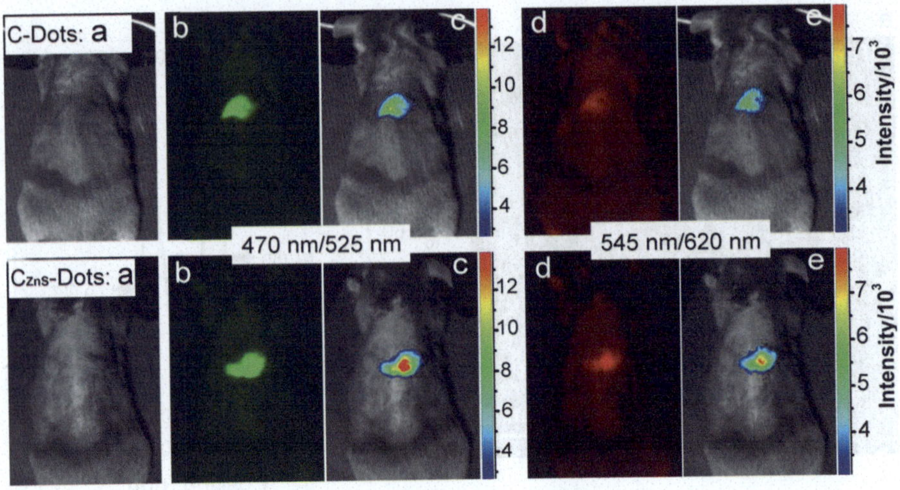

Fig. 6.11 Subcutaneous injection: (**a**) bright field, (**b, d**) as-detected fluorescence (excitation/emission wavelengths indicated), and (**c, e**) color-coded images (ImageJ from NIH). (Originally Figure 2 in Yang et al. (2009a). Reproduced with permission)

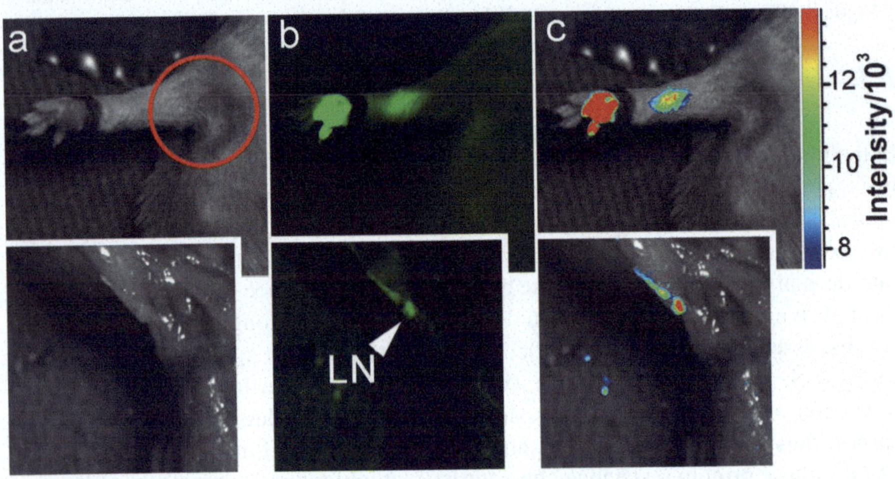

Fig. 6.12 Interdermal injection: (**a**) bright field, (**b**) as-detected fluorescence, and (**c**) color-coded images. Insets: the dissected (in the circled area) axillary lymph node (**LN**). (Originally Figure 3 in Yang et al. (2009a). Reproduced with permission)

which was brighter in the former (Fig. 6.13), consistent with the urine excretion pathway. The relatively weak fluorescence in the dissected liver was an indication of a low accumulation level of the dots. Collectively the results provided strong evidence for the relatively fast clearance of the CDots from mice, not too much longer than 4 h, consistent with the estimated 6 h in a separate study (Yang et al. 2009c).

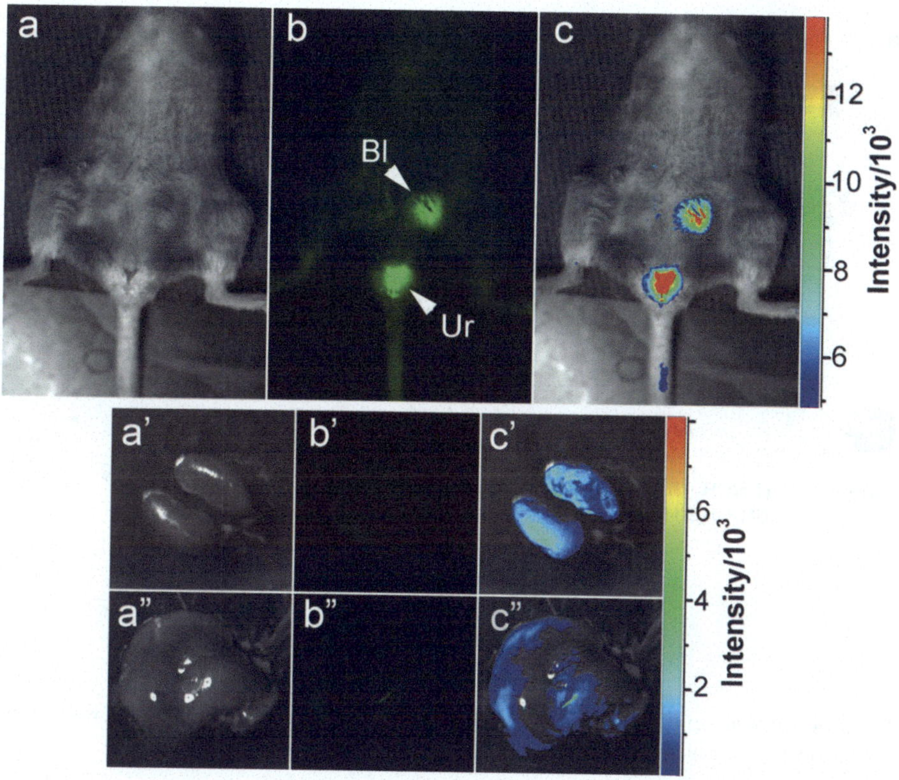

Fig. 6.13 Intravenous injection: (**a**) bright field, (**b**) as-detected fluorescence (**Bl**: bladder and **Ur**: urine), and (**c**) color-coded images. The same order for the images of the dissected kidneys (lower left) and liver (lower right). (Originally Figure 4 in Yang et al. (2009a). Reproduced with permission)

In the study a few years later (Huang et al. 2013), PEG_{1500N}-CDots bearing surface amino moieties (due to PEG_{1500N} being a diamine) were coupled with near-IR dye ZW800 (Fig. 6.14), and the dye-labeled CDots enabled the in vivo imaging in the near-IR spectral region. Effect of three different injection routes (subcutaneous, intramuscular, and intravenous) on the fate of the CDots in vivo was explored by using both near-IR fluorescence (NIRF) and positron emission tomography (PET) imaging techniques. It was found that the CDots were efficiently and rapidly excreted from the body after all three injection routes. The clearance rate of the CDots was found to be intravenous > intramuscular > subcutaneous. The dots had relatively low retention in the reticuloendothelial system (RES) and showed high tumor-to-background contrast. Furthermore, different injection routes for the CDots also resulted in different blood clearance patterns and tumor uptakes of the dots.

On the biodistribution of the CDots by three injection routes (Fig. 6.15), majority of the CDots were accumulated in the kidneys at 1 h after injection via all three routes, and only small amount went to the liver (Huang et al. 2013). The results

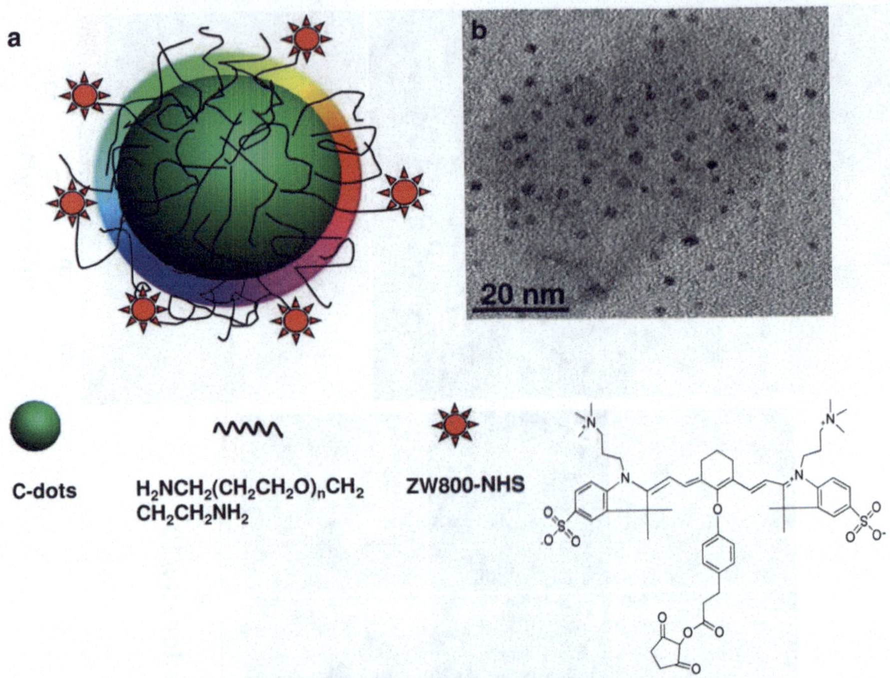

Fig. 6.14 (**a**) Schematic structure of the CDot – ZW800 conjugate. (**b**) TEM image of the CDots. (Originally Figure 1 in Huang et al. (2013). Reproduced with permission)

were rather similar to those reported earlier on the parent PEG_{1500N}-CDots without the dye label (Yang et al. 2009a). After 24 h, no significant signal was detected in any organ, suggesting that the dots were all cleared out of the body regardless of the injection route (Huang et al. 2013). Consistent with these observations were the results from the positron emission tomography (PET) imaging designed for visualizing and quantifying the in vivo kinetics of the CDots. The conclusion based on all these results was that for all three injection routes there was low accumulation of the CDots in the reticuloendothelial system (RES) and rapid clearance from the body mainly via the renal route into urine (Huang et al. 2013).

For the evaluation on the tumor uptake of the CDots (Huang et al. 2013), the dots were injected via three injection routes into athymic nude mice bearing subcutaneous SCC7 tumors. At 2 h post-injection, tumor was easily distinguishable from surrounding normal tissue in all three groups and the tumor signal remained high over time (Fig. 6.16). Most of the remaining dots were excreted through renal filtration, with low level of fluorescence in other tissues and organs. The imaging results were further confirmed by ex vivo microscopic imaging of the frozen tissue slices obtained from mice sacrificed at 2 h post-injection (Fig. 6.16). Co-localization of the CDots and ZW800 in the tissue slices again excluded the possibility of any dissociation of the labeling dye ZW800 from the ZW800-labeled PEG_{1500N}-CDots.

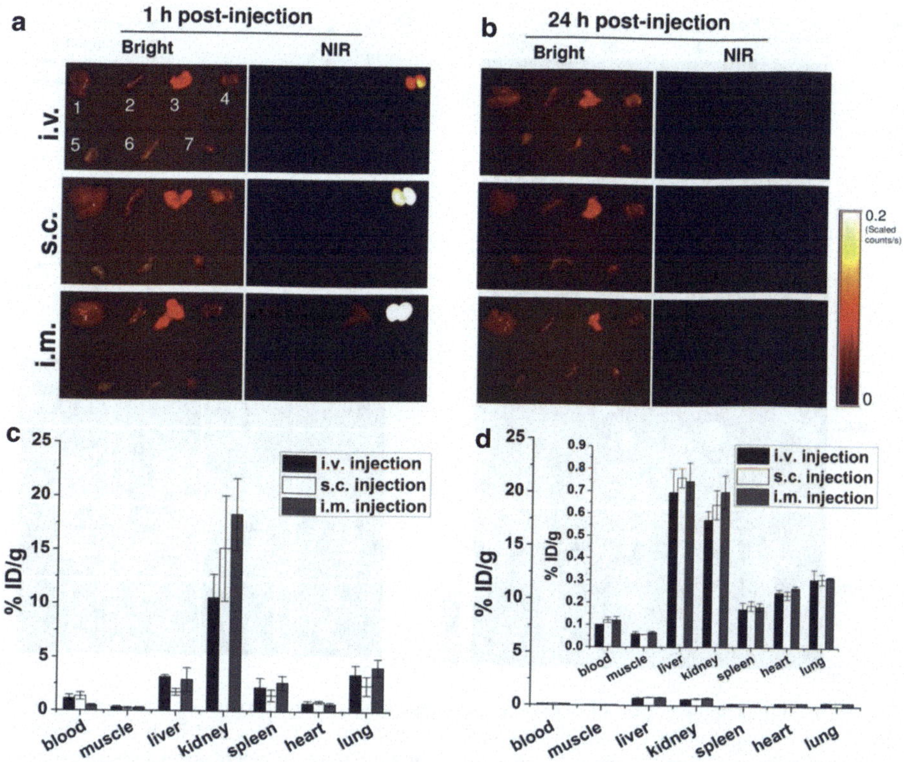

Fig. 6.15 Biodistribution of the dots after three injection routes. (**a**) Ex vivo imaging of biodistribution of the dots. After injection of the dots, the major organs and tissues were harvested from Balb/C mice at the indicated time points, and subsequently, the bright field and NIR imaging were acquired using a Maestro imaging system. Left, 1 h post-injection (top, i.v.; middle, s.c.; bottom, i.m.); right, 24 h post-injection (top, i.v.; middle, s.c.; bottom, i.m.). Bright field: 1, liver; 2, spleen; 3, lung; 4, kidneys; 5, muscle; 6, intestine; 7, heart. (B-C) Quantification of the biodistribution of ^{64}Cu-labeled CDots via three injection routes at 1 h (**b**) and 24 h (**c**) time points. At 1 h and 24 h post-injection of ^{64}Cu-CDot, the mice were euthanized and major organs and tissues including the blood, muscle, liver, kidneys, spleen, heart, and lung were harvested and measured by gamma counting. (Originally Figure 4 in Huang et al. (2013). Reproduced with permission)

The use of near-IR dye ZW800 to label PEG_{1500N}-CDots was an excellent idea to enable in vivo imaging in the more tissue transparent spectral window, because CDots are generally weak absorbers and emitters in the red/near-IR due to the intrinsic spectral coverage of small carbon nanoparticles. In fact, the π-plasmons and associated electronic transitions in carbon nanoparticles are extremely robust in terms of resisting any alterations, making it virtually impossible to prepare any genuine red/near-IR CDots. Many of the reported “red/near-IR carbon quantum dots” from carbonization syntheses for bioimaging studies were simply poorly formed dot structures contaminated or even dominated by dye-like species/mixtures produced in the carbonization processing (Liang et al. 2019). Such materials may

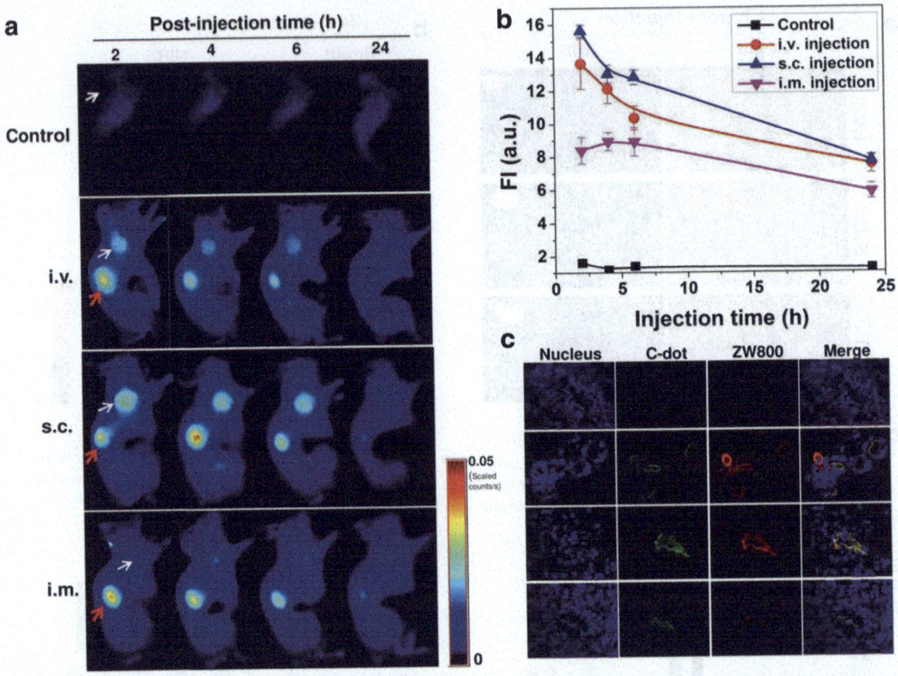

Fig. 6.16 Tumor uptake of CDot – ZW800 conjugate after different routes of injection. (**a**) NIR fluorescence images of SCC-7 tumor-bearing mice acquired at 2, 4, 6, and 24 h post-injection: control (without injection), i.v. injection; s.c. injection; i.m. injection (white arrow indicates tumor; red arrow indicates kidney). (**b**) Tumor ROI analysis. Fluorescence signal unit: $\times 10^8$ photons/ cm^2/s. (**c**) Ex vivo fluorescence images derived from the emission of CDots and ZW800 were acquired to confirm tumor uptake of particles. (Originally Figure 7 in Huang et al. (2013). Reproduced with permission)

conceptually be similar to the ZW800-CDots conjugates in (Huang et al. 2013), but structurally and composition-wise complicated and messy, casting serious doubts on their derived bioimaging results and conclusions. In fact, because dyes are responsible for the imaging results anyway, it is much cleaner and unambiguous to label CDots with red/near-IR dyes in conjugates, as demonstrated in (Huang et al. 2013), or to purposely add such dyes in the dot synthesis for their becoming guest in the resulting host-guest CDots (Sun et al. 2015) for bioimaging applications (see more details in Chap. 7).

2 Ultrabright Fluorescence Probes

CDots are generally compact, or ultracompact in the case of EDA-CDots and EPA-CDots (Fig. 4.1), so they may be used as "building blocks" for larger yet still compact probes of much brighter fluorescence at the individual probe level. As reported

(Anilkumar et al. 2013), CDots could be clustered with a crosslinking agent into probes of fluorescence brightness proportional to the number of dots in the clusters. Dimethyl pimelimidate (DMP), an agent well documented for the crosslinking of proteins, was used to crosslink PEG_{1500N}-CDots into probes of various sizes (Fig. 6.17).

The crosslinked sample was found to be stable against treatments that would otherwise be subject to the de-functionalization effect. "The NMR results were consistent with an expected structure of the DMP-crosslinked PEG_{1500N} functionalization layer as a corona-like soft shell on the surface of a carbon nanoparticle in a single carbon dot, or several such nanostructures crosslinked in a larger probe" (Anilkumar et al. 2013). A mixture of probes in different sizes was analyzed by transmission electron microscopy (TEM) imaging, from which the results suggested the presence of probes containing up to seven carbon nanoparticle cores per probe (Fig. 6.18). The overall image sizes for the probes were up to 25 nm in diameter (or the average edge to edge dimension), though the hydrodynamic dimensions of the probes were probably somewhat larger, as the soft materials at the probe edges (functionalization and crosslinking agents) could not be accurately defined in the TEM imaging. The same sample was analyzed by atomic force microcopy (AFM), and the results were consistent with those from TEM, also suggesting that the sample was a mixture of probes in different sizes (Fig. 6.19).

According to results from the confocal fluorescence imaging of the individual probes on a solid substrate (Fig. 6.20), the probes exhibited different image sizes (submicron in appearance due to the intrinsic resolution limitation of confocal fluorescence microscopy) and varying fluorescence intensities. More quantitatively,

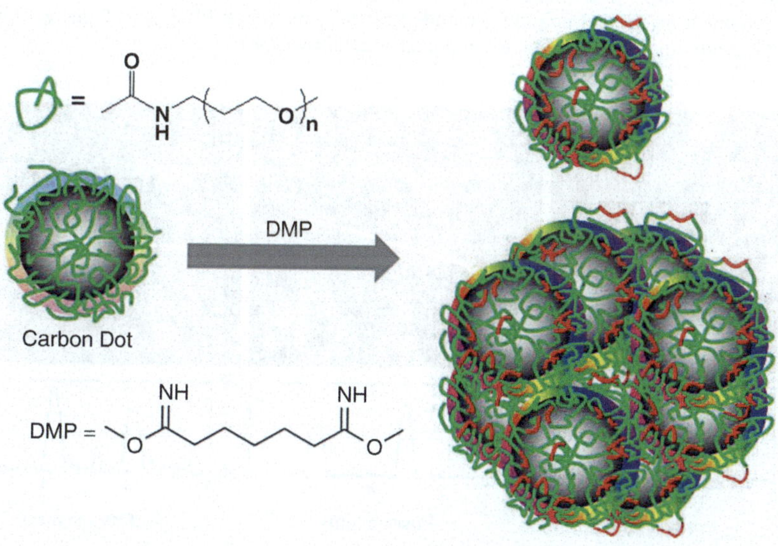

Fig. 6.17 PEG_{1500N}-CDots and crosslinking. (Originally Scheme 1 in Anilkumar et al. (2013). Reproduced with permission)

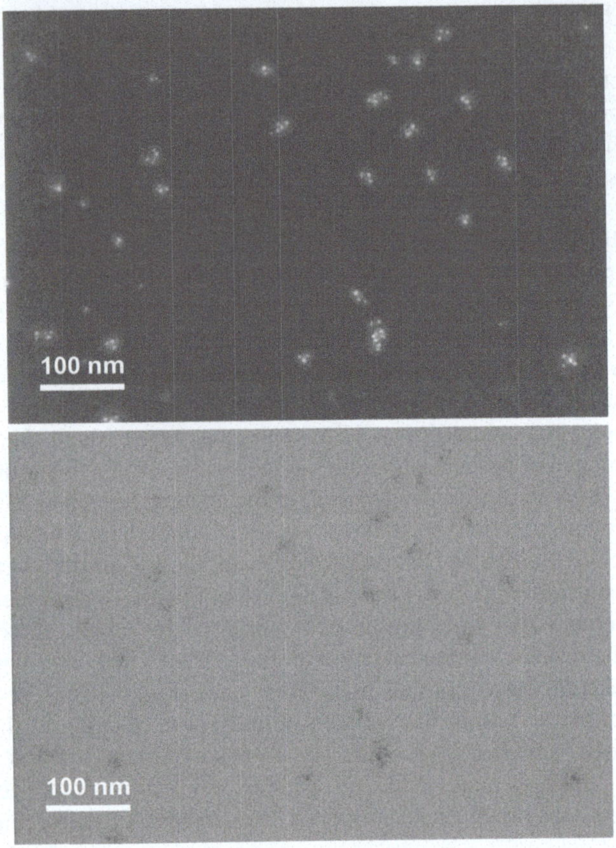

Fig. 6.18 TEM images in Z-contrast (upper) and transmission (lower) modes for the highly fluorescent sample from the gel column separation of the crosslinked PEG_{1500N}-CDots. (Originally Figure 2 in Anilkumar et al. (2013). Reproduced with permission)

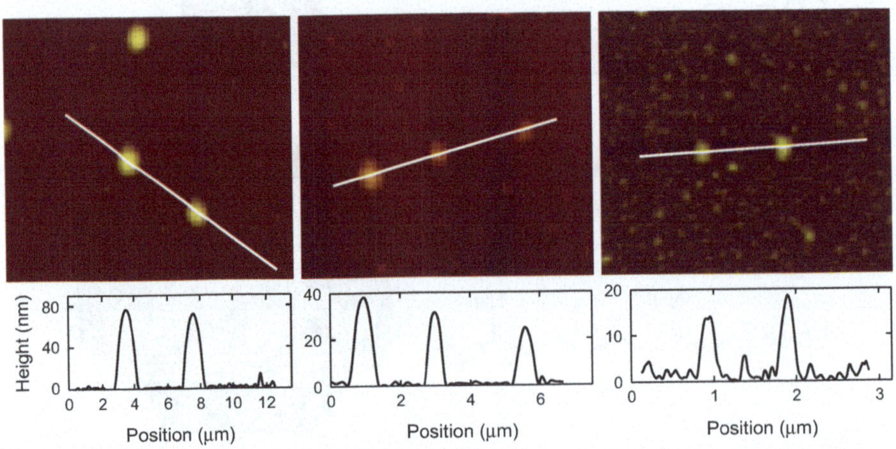

Fig. 6.19 AFM images for representative crosslinked PEG_{1500N}-CDots. (Originally Figure 3 in Anilkumar et al. (2013). Reproduced with permission)

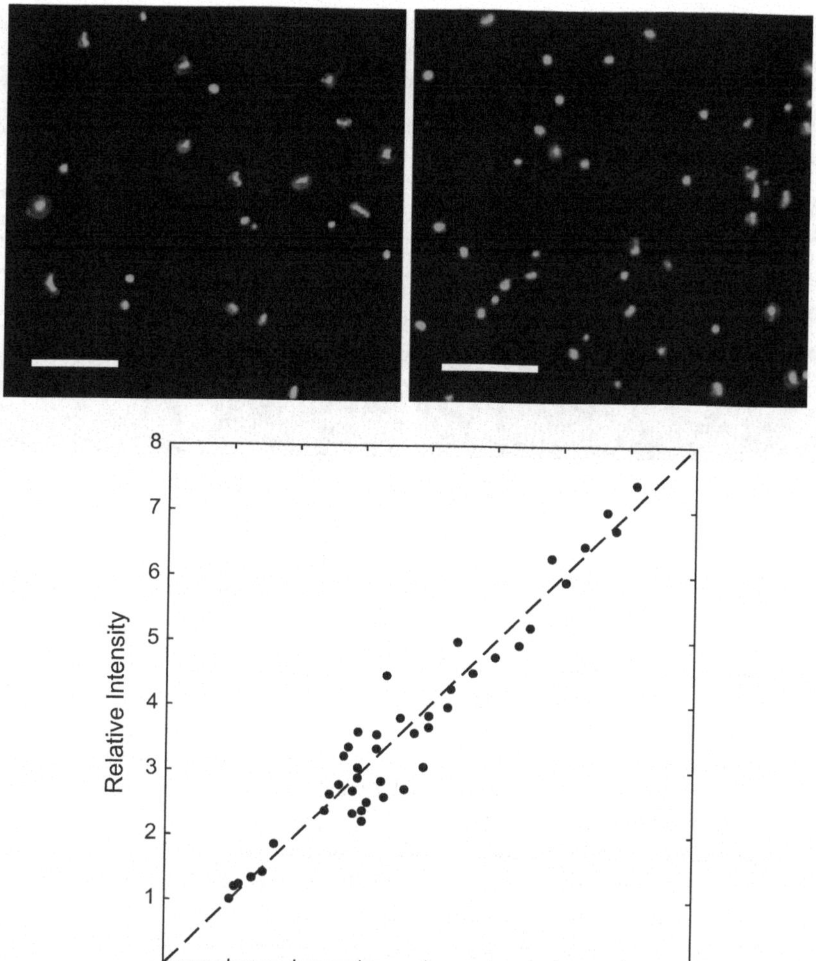

Fig. 6.20 Confocal fluorescence imaging results (upper, scale bars 10 μm) and a correlation between the relative image sizes and integrated emission intensities for the highly fluorescent sample from the gel column separation of the crosslinked PEG_{1500N}-CDots. (Originally Figure 4 in Anilkumar et al. (2013). Reproduced with permission)

relative sizes of the individual probe images were correlated with their corresponding integrated total fluorescence emission intensities (Fig. 6.20). By designating the smallest ones as probes containing only a single dot, the largest probes of about seven times larger in the image size were apparently about seven times brighter in the fluorescence emission. "The correlation between the observed fluorescence image sizes for the individual probes and their actual sizes (varied due to their containing different numbers of crosslinked carbon dots) was confirmed by concurrent

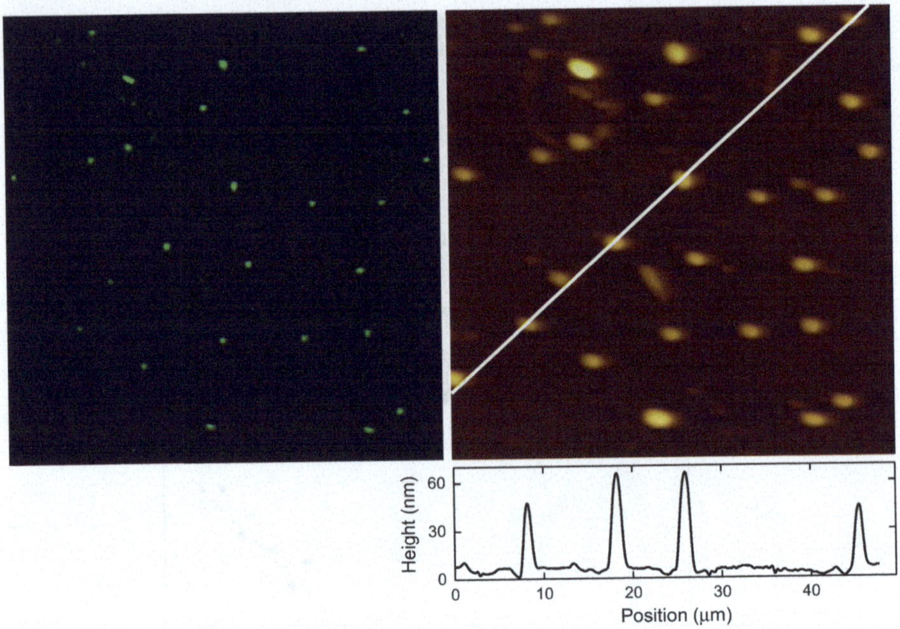

Fig. 6.21 Results from spatially matching imaging of the same specimen by confocal fluorescence microscopy (left) and AFM (right). (Originally Figure 5 in Anilkumar et al. (2013). Reproduced with permission)

confocal fluorescence imaging and AFM measurements of the same specimen" (Fig. 6.21) (Anilkumar et al. 2013).

It was concluded that "the largest probes packed with seven or so carbon dots (Fig. 6.18) could apparently deliver enhanced fluorescence brightness at the individual probe level about seven times higher than that by a single carbon dot (Fig. 6.20). Despite the approximate nature of such an estimation, the results do suggest that crosslinking multiple dots represents an effective way to fabricate ultra-bright fluorescence probes for high-sensitivity cellular imaging and other applications" (Anilkumar et al. 2013). Indeed, for the cubic relationship between the diameter of the probe and the number of dots that could be accommodated in the probe, the packing of a large number of CDots, especially ultracompact CDots, into a single cluster may be explored for extremely bright individual fluorescence probes.

3 Photodynamic Effects

Similar to and competing with conventional semiconductor nanoparticles and QDs in photoexcited state properties and processes, CDots have been studied as photosensitizers for light-activated killing of cancer cells and inhibiting of microbials. Mechanistically, these actions may be associated with two "states" of CDots following photoexcitation, the initially formed charge separated state with electrons and holes trapped at surface defect sites (Fig. 4.7) and the emissive excited state from radiative recombination of the electrons and holes. Thus, the broadly defined photosensitization for photodynamic effects may be explored for a correlation with the mechanistic origins, and conversely the correlation may enable more rational design and manipulation of the dot structures and properties for improved anticancer and antimicrobial outcomes.

The effective harvesting of visible photons by CDots and the significant overlap of their optical absorptions with the solar spectrum (Fig. 2.20) make the dots readily activated by visible/natural light, a major advantage over the widely used nanoscale semiconductors such as colloidal TiO_2 that require UV excitation. Carbon nanoparticles as the dominating ingredient in CDots are abundant and can also be produced from renewable sources, and dot samples can be made to be very inexpensive and in bulk quantities, so that CDots will be competitive in performance and costs for broad applications.

3.1 Anticancers

As summarized in (Juzenas et al. 2008), CDots could be a competitive platform in the broad field of quantum dots and nanoparticles for photodynamic and radiation therapies of cancers. Indeed, there have been studies to explore and demonstrate the potential of CDots as anticancer agents in both photodynamic and radiation therapies. For example, PPEI-EI-CDots (Fig. 4.1) were used as photosensitizer or photocatalyst for enhanced photodynamic effects on the survival of human prostate adenocarcinoma Du145 and PC3 cells (Fig. 6.22) (Juzenas et al. 2013).

CDots could be surface-doped with noble metals like silver in a relatively simple photo-reductive deposition experiment, as discussed in Chap. 4. The silver metal-doped PEG_{1500N}-CDots were explored as a bimodal sensitization platform for both photodynamic therapy (PDT) and radiotherapy (RT) functions, with the latter taking advantage of the large cross-section for X-ray (Kleinauskas et al. 2013). For the survival of human prostate adenocarcinoma Du145 cells post-treatment with X-ray, the presence of silver metal-doped PEG_{1500N}-CDots resulted in significantly larger reductions (Fig. 6.23).

3.2 Antimicrobials

There have been many reports in the literature on CDots inhibiting various bacteria. The reported observations may be divided into several categories, including those with deliberate light activation, “accidental” light exposure under ambient conditions that are not purposely kept dark, and dot samples carrying or made from known antibiotics or disinfection agents. For the last category, CDots essentially

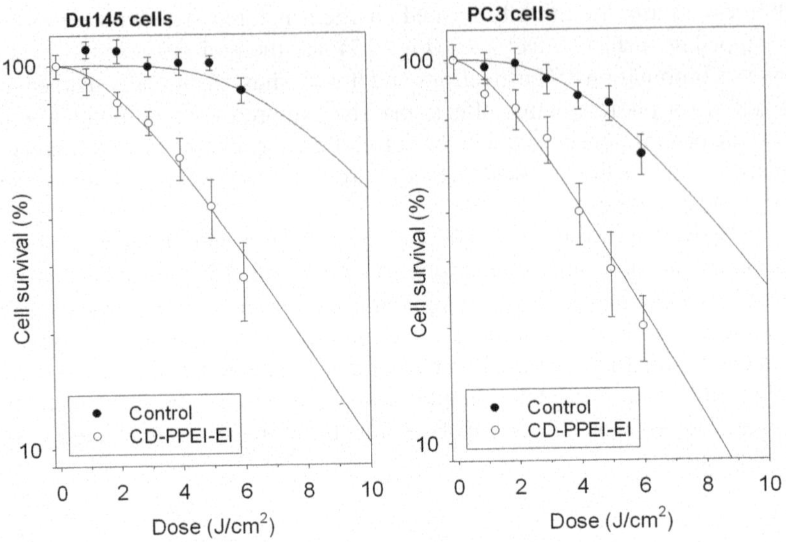

Fig. 6.22 Survival of the human prostate adenocarcinoma Du145 and PC3 cells post-treatment by photoirradiation without (control) and with PPEI-EI-CDots. For the latter, the cells were incubated with the CDots for 24 h prior to the photoirradiation. (Originally Figure 4 in Juzenas et al. (2013). Reproduced with permission)

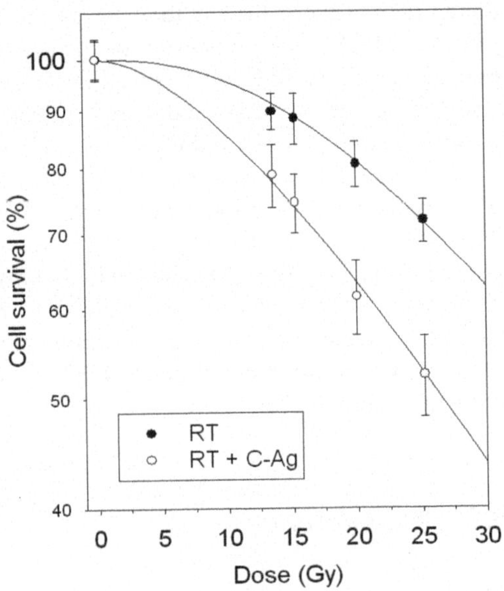

Fig. 6.23 Survival of the human prostate adenocarcinoma Du145 cells post-treatment by 160 kV X-ray (dose rate 1 Gy/min) without (control) and with the silver metal-doped PEG_{1500N}-CDots. For the latter, the cells were incubated with the dot sample for 20 h prior to the irradiation treatment. (Originally Figure 8 in Kleinauskas et al. (2013). Reproduced with permission)

serve the drug delivery function, not due to any photoinduced properties or processes of the dots. On the light-activated antibacterial activities, their significant inhibition effect even under ambient light conditions demonstrates the high effectiveness of CDots as light-activated bactericidal agents.

In the study reported in (Meziani et al. 2016), antibacterial activities of CDots were observed under both ambient light and deliberate visible light exposure conditions. As shown in Fig. 6.24, EDA-CDots could inhibit *E. coli* cells substantially with just exposure to ambient light in a safety cabinet for an hour.

In a more quantitative evaluation, a light box equipped with a 36 W light bulb was used for visible light illumination. The samples of *E. coli* with and without (as control) EDA-CDots were treated in the light box or in the dark for 30 min. According to the measured viable cell numbers (Fig. 6.25), the dot treatment coupled with visible light illumination was obviously effective for bacteria killing, with about 4 logs of *E. coli* cells killed, significantly more effective than the outcome with only ambient light exposure (Fig. 6.24) (Meziani et al. 2016). Nevertheless, the results were all consistent in terms of confirming the visible/natural light-activated bactericidal function of CDots.

In the same study (Meziani et al. 2016), a different experimental configuration was employed for evaluating the antibacterial activities of CDots on surface. Bacteria cells were mixed with EDA-CDots and plated on trypticase soy agar (TSA) plates. Upon the exposure of the plates to visible light, there were obvious differences between the treated plates and the controls (Fig. 6.26), again suggesting substantial bactericidal effect of CDots with visible light illumination.

The bacteria killing was attributed generally to the photodynamic effect of CDots under visible light for the generation of reactive oxygen species (ROS), both intracellular and extracellular, to cause lipid peroxidation and membrane damages (Fig. 6.27) (Dong et al. 2020). Such a mechanism has been supported by experiments to capture the intracellular ROS and to probe the damages directly (Abu Rabe et al. 2019).

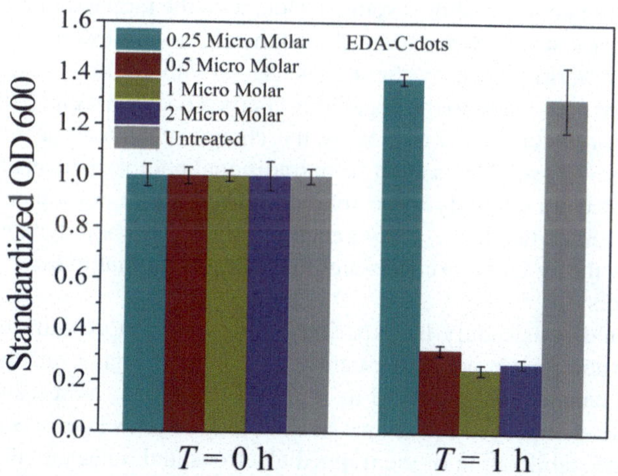

Fig. 6.24 Photoinduced bacteria inactivation measured by the optical density at 600 nm (OD600) for EDA-CDots (presented as mean ± standard deviation of quadruplicate experimental results). (Originally Figure 4 in Meziani et al. (2016). Reproduced with permission)

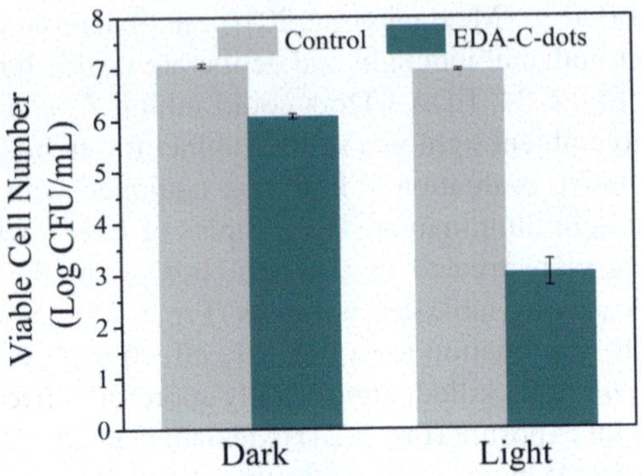

Fig. 6.25 The reductions in the viable cell number after *E. coli* cells were treated with EDA-CDots for 30 min with or without light (presented as mean ± standard deviation of triplicate experimental results, and the data were also analyzed using Student *t* test with $P < 0.05$ as significant difference). (Originally Figure 6 in Meziani et al. (2016). Reproduced with permission)

On photoinduced ROS generation and actions in the mechanistic framework of CDots (Fig. 4.7), relative contributions from the surface-trapped electrons and holes and the emissive excited states should be considered. Upon photoexcitation the initial charge separation for the formation and trapping of electrons and holes is on ultrafast (picosecond or even sub-picosecond) time scale, followed by similarly fast radiative recombinations (as made evidence by the ultrashort fluorescence rising time) for the emissive excited states, whose decays are relatively much slower on nanosecond time scale. The surface-trapped electrons and holes are themselves ROS-like, in addition to their contributions to the classical ROS generation, and their related antimicrobial activities are on the fast time scale, likely requiring near-neighbor or close contacts of the excited CDots with the targeted microbial species. Relevant experimental evidence has included the observed substantial improvement in antibacterial performance by the use of CDots with positively charged surface moieties, which could bind to the negatively charged bacterial cells (Abu Rabe et al. 2019). These antimicrobial activities by the charge separated species on the fast time scale (Fig. 4.7) set CDots apart from traditional photosensitizers that are commonly employed for photodynamic inhibition of bacteria and other microorganisms. On the other hand, the ROS generation and the photodynamic inhibition contributed by the emissive excited states in CDots are similar to those of traditional photosensitizers.

As suggested consistently by experimental results, CDots with more effective surface passivation are more fluorescent with observed higher quantum yields. In the same mechanistic framework (Fig. 4.7), the enhanced fluorescence emissions (larger Φ_F) due to improved surface passivation may be associated more or predominantly with the stabilization of the trapped electrons and holes and their more productive radiative recombinations (collectively reflected in Φ_1), because according to the consistently obtained fluorescence decay results the emissive excited states (with the slower processes represented by Φ_2) are mostly insensitive to the effective-

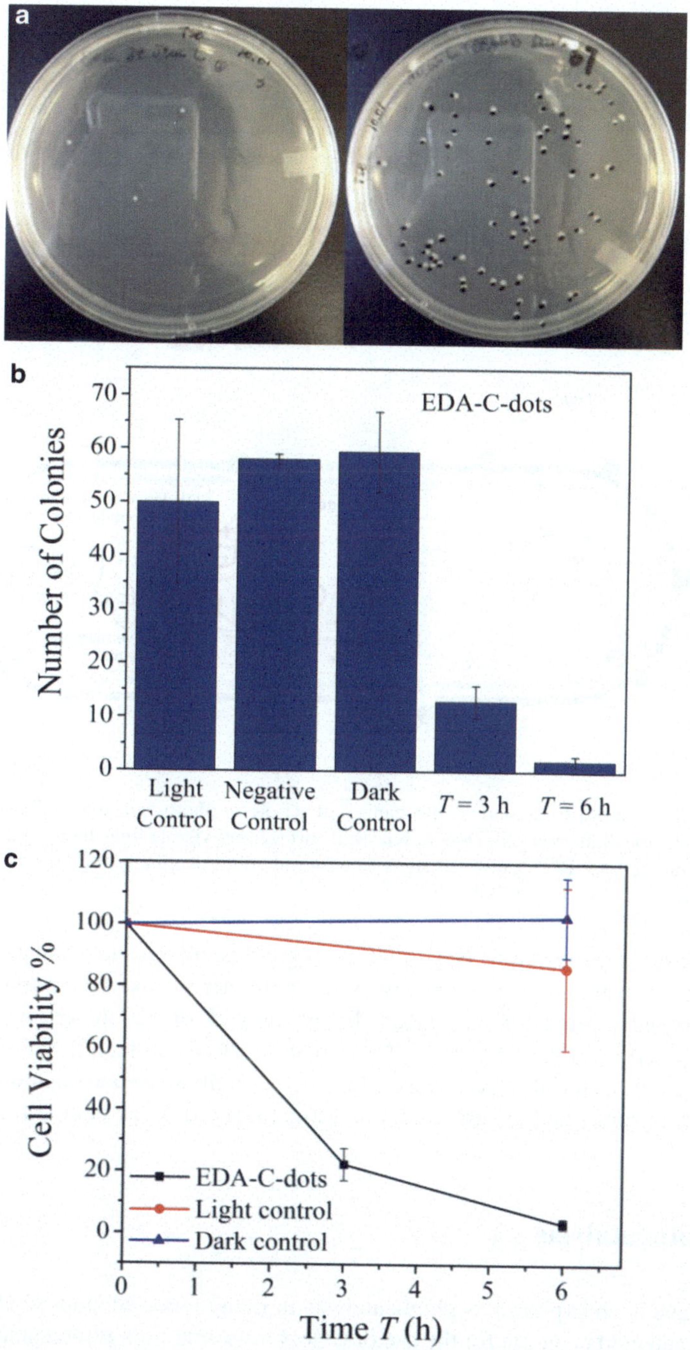

Fig. 6.26 Photoinduced bacteria inactivation by EDA-CDots: (**a**) Photos showing colonies on TSA plates (left: 6 h treatment; right: dark control); (**b**) *E. coli* colony numbers; and (**c**) Cell viability ratio. The CDots were presented in the dark control. (Originally Figure 7 in Meziani et al. (2016). Reproduced with permission)

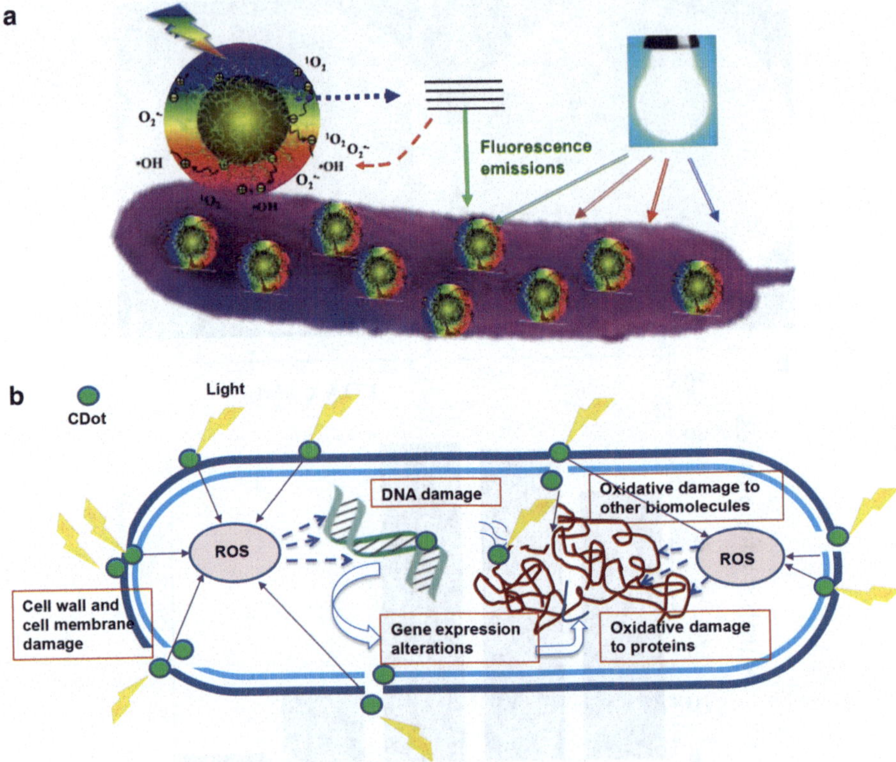

Fig. 6.27 Cartoon illustration on the mechanism of action for photo-activated antibacterial activities of CDots. (**a**) Adhesion of CDots to bacterial surface and visible light-induced generation of ROS. (**b**) Intracellular ROS cause damages to bacterial cell. (Originally Figure 6 in Dong et al. (2020). Reproduced with permission)

ness in surface passivation. Nevertheless, regardless of the mechanistic details it was correct to suggest that "CDots with the more effective surface passivation and correspondingly more fluorescent are likely more desirable in serving as visible light-activated bactericidal agents" (Meziani et al. 2016). In fact, the suggested correlation of antibacterial effectiveness of CDots with fluorescence quantum yields of the dots was confirmed in subsequent investigations (Al Awak et al. 2017).

4 Photocatalysis

CDots have been explored as photocatalysts in many reported studies. Highlighted here is a selected example for the use of CDots in visible light photocatalytic reduction of CO_2 into small organic molecules, with the selection based on the fact that the CO_2 conversion efficiencies thus achieved are among the best in the relevant field (Cao et al. 2011a; Sahu et al. 2014).

As first reported (Cao et al. 2011a), the PEG_{1500N}-CDots with surface-doped gold or platinum metal were used as photocatalysts for the reduction of CO_2, where the gold or platinum metal was for concentrating electrons to diminish fluorescence emissions yet enhance substantially the photocatalytic reduction (Fig. 6.28). Experimentally, the aqueous dot solution saturated with CO_2 was photoirradiated with visible light (425–720 nm) for 5 h. Formic acid as a volatile photoproduct in the aqueous solution post photoirradiation was harvested for identification and quantification. The formation of formic acid as a photoproduct was confirmed in ^{13}C-labeling experiments by using aqueous $NaH^{13}CO_3$ (at the saturated CO_2-equivalent concentration) as the feed under neutral and acidic pH conditions (thus mimicking aqueous CO_2), so as to enable unambiguous ^{13}C-NMR characterization.

The quantification of the photoproduct was used to estimate quantum yields for the photocatalytic reduction of CO_2. For the reduction to formic acid only, the estimated quantum yield was about 0.3%, which was likely lower than the overall quantum yield for the CO_2 conversion, as it was generally known that the photoconversion would yield other products such as methanol (Cao et al. 2011a).

In subsequent investigations (e.g., Sahu et al. 2014), the same CDots with gold doping in aqueous solution were used with CO_2 at higher concentrations in a sealed optical cell for the same visible light (405–720 nm) photocatalytic reduction (Fig. 6.29). In addition to the substantial presence of formic acid, acetic acid was also detected in the product mixture, which was rather surprising because the reduction to acetic acid requires overall eight electrons, regardless of whichever detailed mechanisms debated in the literature.

Formic acid as a prominent photoproduct was again used to determine the relative conversion efficiencies at different CO_2 pressures (corresponding to different CO_2 concentrations in the aqueous solution). The formic acid production obviously increased with the increasing CO_2 pressure (or concentration, Fig. 6.30). At 1900 psia, the amount of formic acid was about an order of magnitude higher than that produced under ambient CO_2 pressure, which should correspond roughly to the

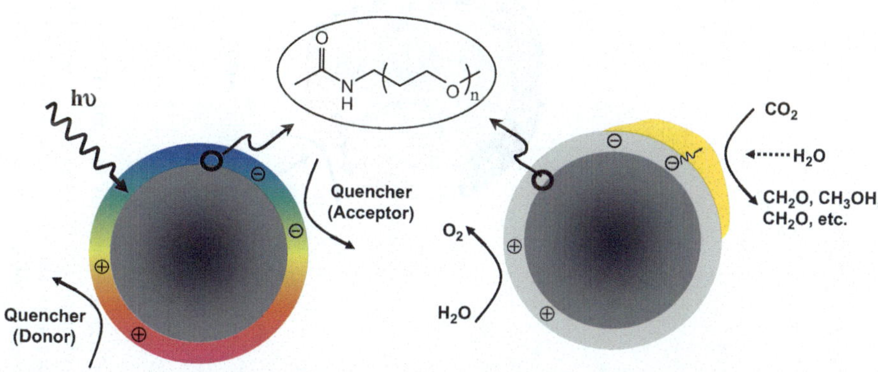

Fig. 6.28 Aqueous soluble PEG_{1500N}-CDots before (left, fluorescent) and after (right, photocatalytic for CO_2 reduction in aqueous solution) the metal coating. (Originally Figure 1 in Cao et al. (2011a). Reproduced with permission)

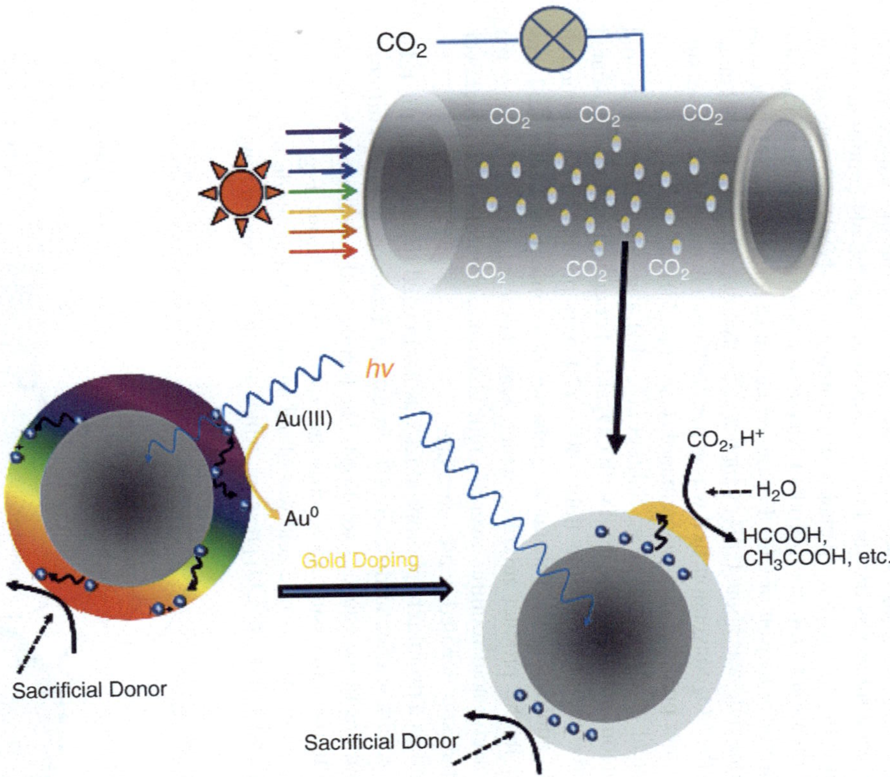

Fig. 6.29 Cartoon illustrations on (upper) the high-pressure optical reactor; (lower-left) the photoreductive doping of the CDot with gold, completely quenching the dot surface-based fluorescence (illustrated as the change of the dot surface from rainbow to grey); and (lower-right) the gold-doped CDot as photocatalyst for CO_2 conversion, where the doped gold (in yellow) was small in quantity, insufficient to form a shell, and likely random in terms of size and shape. The sacrificial electron donor was isopropanol added in some experiments or PEG_{1500N} molecules on the surface of CDots in the absence of isopropanol. (Originally Figure 2 in Sahu et al. (2014). Reproduced with permission)

same magnitude of increase in the photocatalytic reaction quantum yield for the formic acid formation alone (around 0.3% in the CO_2-saturated aqueous solution under ambient pressure (Cao et al. 2011a), as discussed above). The strong dependence on CO_2 pressure (equivalent to CO_2 concentration) was really unexpected, "which might be understood in terms of the importance of the initial interactions between the photoexcited catalysts and CO_2 molecules in the conversion to formic acid, with more CO_2 molecules in the aqueous solution under a higher CO_2 pressure in the optical cell, and/or the involvement of more complex multi-step processes in the photoconversion reaction" (Sahu et al. 2014).

The results highlighted above and more in the literature have shown that CDots with surface-doped noble metals are unusually effective photocatalysts for CO_2 photoreduction. On the possible mechanistic origins (Cao et al. 2019): "The effectiveness of the CDots-based photocatalysts may be rationalized by the available results and mechanistic discussions in the use of conventional semiconductors as photocatalysts. In fact, most of the more efficient semiconductor photocatalysts for CO_2 conversion have been those constructed with several semiconductors of different bandgaps, with the catalyst design principle and practice reflecting implicitly or explicitly the need for some kind of coordinated actions of multiple catalysts in the

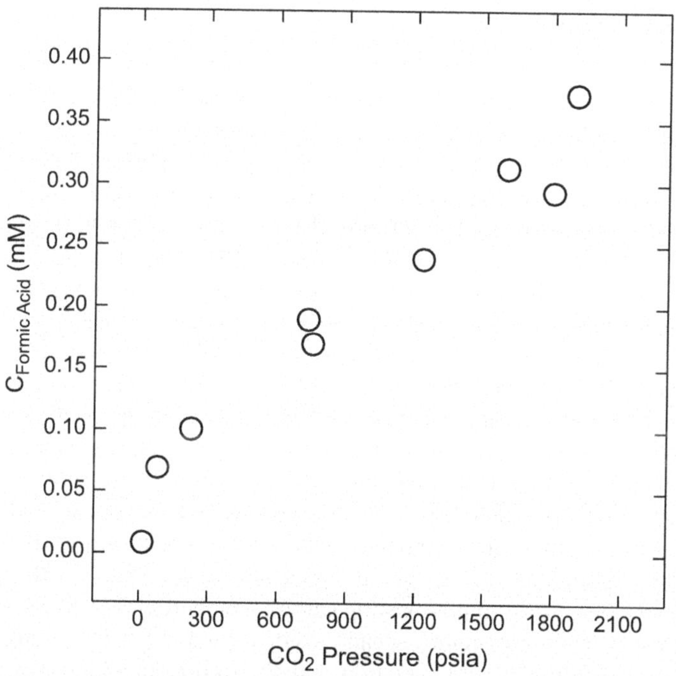

Fig. 6.30 Results in terms of formic acid concentrations in the reaction mixtures from a series of photoconversion reactions in aqueous solution of the CDots under different CO_2 pressures (in the long-path optical cell as the high-pressure reactor). (Originally Figure 4 in Sahu et al. (2014). Reproduced with permission)

CO_2 photoreduction reaction. In this regard, CDots for their broadly distributed electronic transitions and a distribution of photoexcited states in individual dots, as reflected by the observed absorption and fluorescence emission properties in bulk and at the individual dot level, may each represent intrinsically a mixture or distribution of photocatalysts of different characteristics. This is analogous to the catalyst configuration of combining multiple semiconductors in one, as discussed above, which may be responsible for the observed more potent photocatalytic functions of CDots and noble metal-doped CDots."

In the literature, CDots and their derived photocatalysts have actually been more popular in investigations on photocatalytic water-splitting. The same photocatalysts of the CDots with surface-doped noble metals used for the CO_2 conversion discussed above were also capable of splitting water under visible light illumination (Cao et al. 2011a). However, additional more quantitative investigations have suggested consistently that these photocatalysts are much more potent for CO_2 reduction than for water-splitting with visible photons. This interesting divergence has major mechanistic implications, deserving more attention in the further development of CDots-based/ CDots-derived photocatalysts.

5 Optoelectronics

Conventional semiconductor QDs have been widely used in optoelectronic devices, such as now commercially available "QLED" television sets and other "quantum dot displays" or LEDs in general. They are also popular in photovoltaic devices, and so are fullerenes, despite the generally poor photon-harvesting capabilities of the widely used C_{60} derivatives like PCBM ([6,6]-phenyl C61 butyric acid methyl ester). As increasingly recognized (Cao et al. 2019), "the optical properties and photoinduced redox characteristics of CDots resemble in many respects those of conventional nanoscale semiconductors, not only phenomenologically but also to a significant extent mechanistically. In that context CDots may be considered as semiconductor-like. On the other hand, CDots represent a nano-structural configuration that materializes and enhances the intrinsic properties of small carbon nanoparticles through the effective particle surface passivation, rivaling fullerenes in many respects." Thus, CDots are naturally pursued as high-performance competitors and/or lower-cost alternatives to conventional semiconductor QDs and/or fullerenes for uses in various optoelectronic devices (Fernando et al. 2015; Cao et al. 2019). Highlighted here are some representative examples for the device use of dot samples derived from pre-existing carbon nanoparticles or likely with dominating nanoscale carbon domains, because as discussed in Chap. 5 many of the dot samples from insufficient carbonization and their corresponding properties may have to be re-examined in reference to genuine CDots.

"CDots with their bright and colorful fluorescence emissions are obviously more useful as emitting materials in light-emitting diodes (LEDs) than their fullerene counterparts, which are known to be only weakly emissive even with a variety of possible derivatizations of the fullerene cages. In fact, CDots have been evaluated as emissive components in LEDs, with either optical or electrical excitations" (Cao et al. 2019).

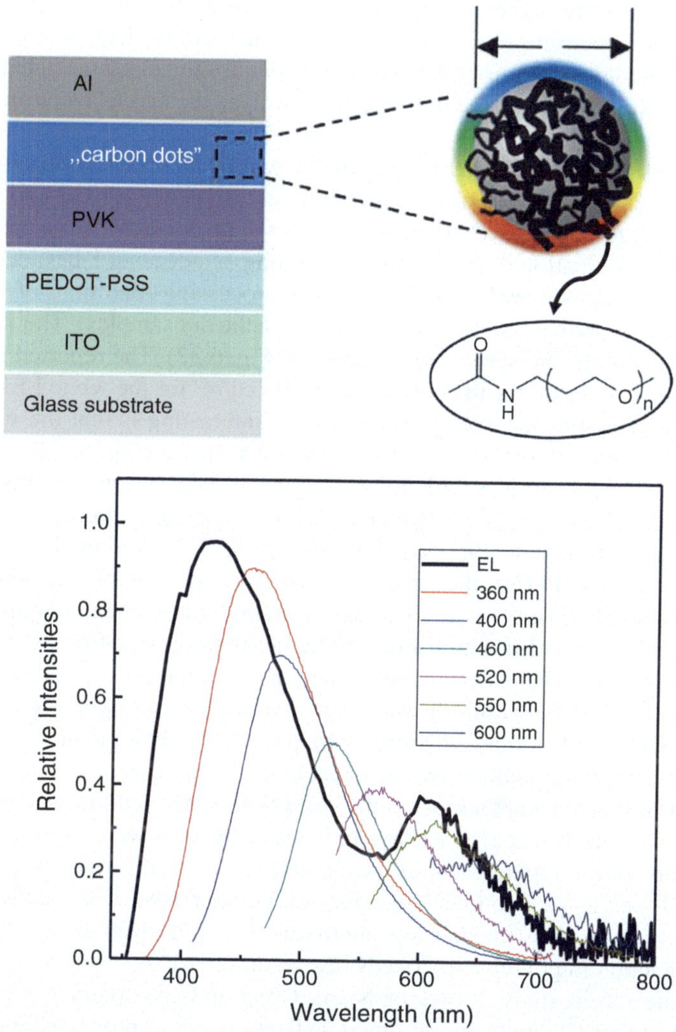

Fig. 6.31 Upper: CDots diode – the architecture of the device with the ITO (anode), PEDOT:PSS (HIL), PVK (HTL), CDots (EL), Al (cathode). Lower: Comparisons between the electroluminescence (EL) spectrum from the LED device and the corresponding fluorescence spectra of the CDots in aqueous solution at the listed excitation wavelengths. (From Veca et al. (2014). Reproduced with permission)

"LEDs with electrical excitation are more advantageous in many respects, without the need for any other LED sources. In the electroluminescent LED architecture, the CDots-based emissive layer is typically sandwiched between the electron and hole transport layers as the most efficient configuration. The fabrication has benefited from the solubility of CDots in common solvent systems, with facile processing at low costs, and good stability and superior efficiency in the resulting LEDs" (Cao et al. 2019). As reported in (Veca et al. 2014), PEG_{1500N}-CDots (Fig. 4.1) were used for the emissive layer. The resulting LED had a good wavelength coverage over the visible spectrum (Fig. 6.31), but the brightness of the LED was not so high, which could be attributed to effects such as residual moisture associated with the highly hydrophilic nature of the dot sample, as well as the lack of electron transport layer in the LED architecture.

In the work reported in (Zhang et al. 2013), two kinds of dot samples were prepared and used for light-emitting diodes, one emitting white light and the other emitting blue light with different applied voltages. The dot syntheses were based on the thermal carbonization of the mixture containing octadecene, 1-hexadecylamine, and citric acid under somewhat more aggressive processing conditions of 300 °C for 3 h (thus likely more nanoscale carbon domains in the dot samples). The fluorescent dots were used in the emissive layer of devices (Fig. 6.32). The reported maximum brightness was 24 cd/m^2 for the blue LED and 96 cd/m^2 for the white LED, and the devices had a low turn-on voltage of 5 V. Also interesting is that the white LED could emit light in different colors under different voltages (Fig. 6.32).

For the use of dot samples with dominating nanoscale carbon contents in photovoltaic devices (Yan et al. 2016; Cui et al. 2018), the samples were prepared using a CVD setup (acetylene as the feed at 1000 °C, Fig. 5.2) (Yan et al. 2016). In the polymer solar cells (PSCs), the purpose of using the dot samples for the electron transporting layer (ETL) was to overcome the limitations of the commonly used ETL materials such as LiF and Ca for the higher thermal stability, enhanced electron injection, and lower interface resistance in the solar cells, as achieved in the study (Yan et al. 2016). Similarly, the CVD produced dot samples were employed in the fabrication of composite films with poly(3-hexylthiophene) (P3HT) (Cui et al. 2018). As made evident by the quenching of fluorescence emissions, there were electron transfer interactions in the composite films with the dots as electron acceptors. The electron acceptor characteristics of the dots were exploited for their combination with or replacement of the C_{60} derivative PCBM in the solar cells, and both were found to be able to enhance the cell performance. It was concluded that CDots could be used as inexpensive alternative to C_{60} derivatives for the electron acceptor function in polymer solar cells (Cui et al. 2018).

In a more recent study, Professor Xiong Gong and coworkers at University of Akron found that CDots from the functionalization of pre-processed and selected small carbon nanoparticles with Jeffamine® D-230 could be incorporated into polycrystalline perovskite solar cells for significantly improved performance.

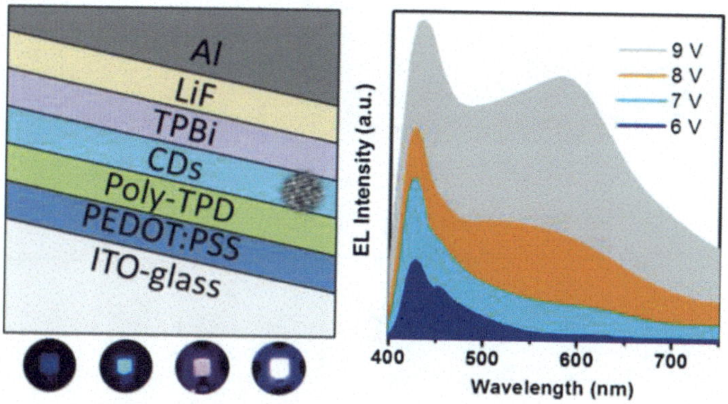

Fig. 6.32 An illustration on the LED device with carbon dots (left) and electroluminescence spectra for different applied voltages. (From Zhang et al. (2013). Reproduced with permission)

6 Concluding Remarks

CDots representing a distinctive carbon nanomaterials platform should have extremely broad applications across many technological fields, for which highlighted here are just some representative examples. A major challenge is and will continue to be with the quality of dot samples, because dye-like species/mixtures crosslinked with some minor nanoscale carbon domains may complicate the understanding of the results and findings important to the specific applications. Many of the conceptually innovative demonstrations in the literature on the excellent application potential of these carbon nanomaterials may need to be re-examined/validated in light of the growing concerns in the research community on dot samples from one-pot carbonization syntheses under poorly selected processing conditions.

Chapter 7
Carbon Hybrid Dots

Nanoscale carbon apparently has such an amazing ability or compatibility in terms of "mixing" (broadly defined) with other elements, molecules, and/or nanomaterials to form carbon-based/carbon-derived hybrid nanostructures or simply "carbon hybrid dots" in a variety of compositions and structural configurations. These hybrid dots represent a major expansion of the CDots platform for nanomaterials properties and related applications that are beyond those of the "parent/neat" CDots. Highlighted here are some representative examples.

1 Host–Guest CDots

As defined, CDots must be built upon small carbon nanoparticles. Logically, the use of pre-existing carbon nanoparticles is more straightforward for the requirement. However, available experimental results suggest that a dot configuration containing a sufficient amount of nanoscale carbon domains, mixed with some organic species (Fig. 5.9), can largely mimic the classical CDots based on pre-processed and selected small carbon nanoparticles in terms of the observed optical properties and photoinduced redox characteristics (Ge et al. 2018), despite the expected structural difference in the dot core (likely more porous carbon–organic mixture in the former (Fig. 5.9) versus solid carbon in the latter (Fig. 1.1)). In the carbonization synthesis for such a dot configuration, the processing conditions must be sufficiently robust to ensure the required substantial presence of nanoscale carbon domains in the resulting dots. One set of such processing conditions is the microwave irradiation of solid-state organic mixtures (Ge et al. 2018), as already highlighted in previous chapters.

© Springer Nature Switzerland AG 2020
Y.-P. Sun, *Carbon Dots*, https://doi.org/10.1007/978-3-030-41184-8_7

1.1 Synthesis and Characterization

The same microwave-assisted processing of solid-state organic mixtures could be applied to the preparation of CDots with encapsulated/embedded organic dye species, for which specifically selected dye molecules were added purposely to the organic precursor mixture for the carbonization (Sun et al. 2015). The rationale for the processing conditions was such that the other organic species were in large excess over the added dyes in the precursor mixture, thus taking much more microwave energy to be carbonized, and the small carbon domains thus created would preferentially absorb more microwave irradiation. As a result, some of the added dyes in the dot structure could survive the rather harsh carbonization processing conditions for the as-designed "host–guest" CDots (denoted as G@CDots, Fig. 7.1,

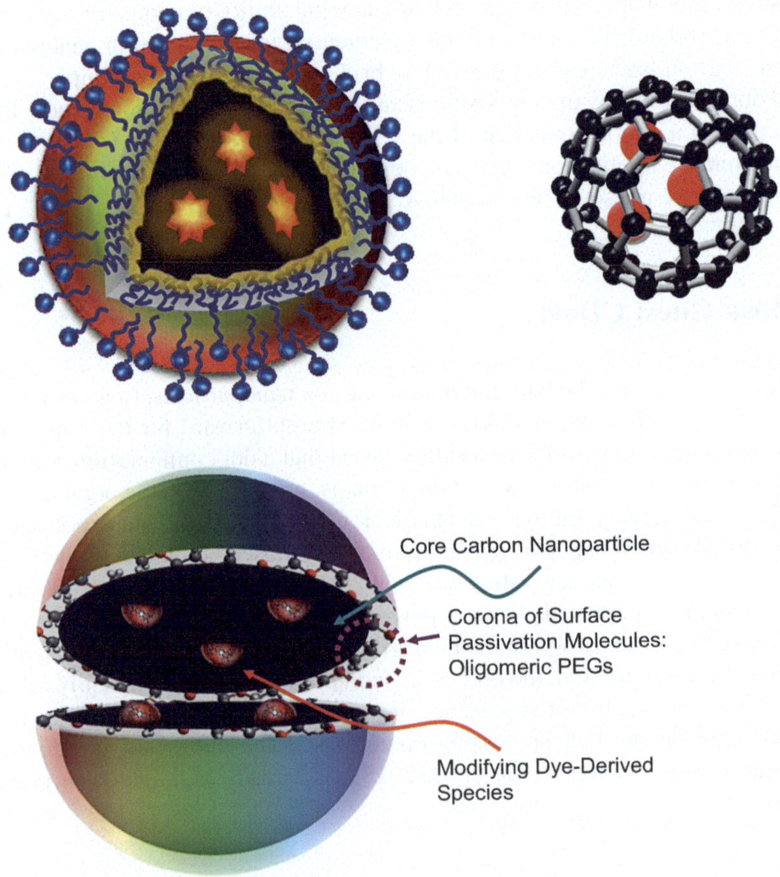

Fig. 7.1 Cartoon illustrations used for (upper) comparison between a host–guest CDot (left) and an endofullerene (right) (Adopted from Sun et al. (2015)); and (lower) the expected structure in a host–guest CDot (Adopted from LeCroy et al. (2019))

with "G" for the encapsulated/embedded dye species as guest). For example, nile blue (NB), cresyl violet (CV), and zinc phthalocyanine (ZnPc) were mixed with a large excess of oligomeric PEGs, and the resulting solid-state mixtures were processed by microwave irradiation to obtain G@CDots (G = NB, CV, and ZnPc) (Sun et al. 2015). Their absorption and fluorescence spectra are shown in Figs. 7.2 and 7.3.

Because of the harsh processing conditions, the guest dye species in G@CDots may or may not be the same as the originally added dye molecules. Therefore, in NB@CDots as an example, the guest should be more accurately referred to as "nile blue-derived dye species." A major objective for the design and preparation of G@CDots has been to move the wavelength coverage of the dots into red and near-IR spectral regions. Apparently, such an objective can be achieved even with the possible or likely alterations/damages to the originally added dye molecules during the carbonization processing, as long as the encapsulated/embedded dye species still preserve strong absorptions and fluorescence emissions in the originally targeted spectral regions (Figs. 7.2 and 7.3). In this regard, there is no real limit in the design and preparation of G@CDots as optical probes for the desired specific wavelength coverage.

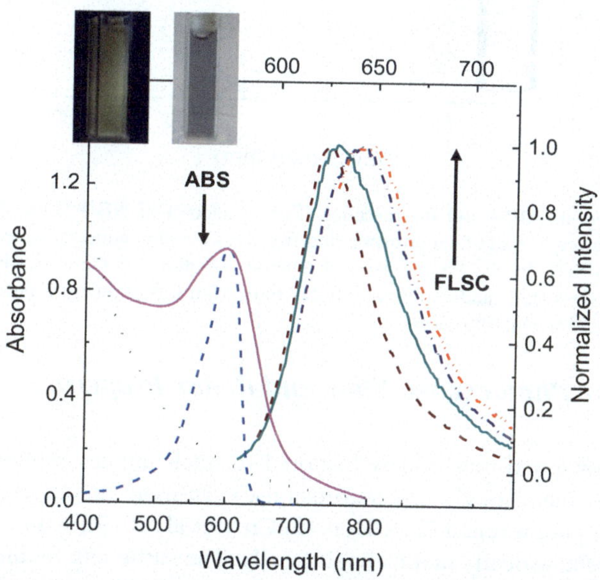

Fig. 7.2 The absorption (ABS) spectrum of CV@CDots (—) and corresponding fluorescence (FLSC) spectra (excitations at 570 nm, —; 600 nm, -.-; and 620 nm, -..-) in aqueous solution. The spectra of free CV (- - -) and CDots from the carbonization of PEG_{900} without any encapsulation (. . .) in aqueous solutions are also shown for comparison. Inset: Photographs of an aqueous solution of the sample under UV light in the dark (left) and under natural day light (right). (Originally Figure 2 in Sun et al. (2015). Reproduced with permission)

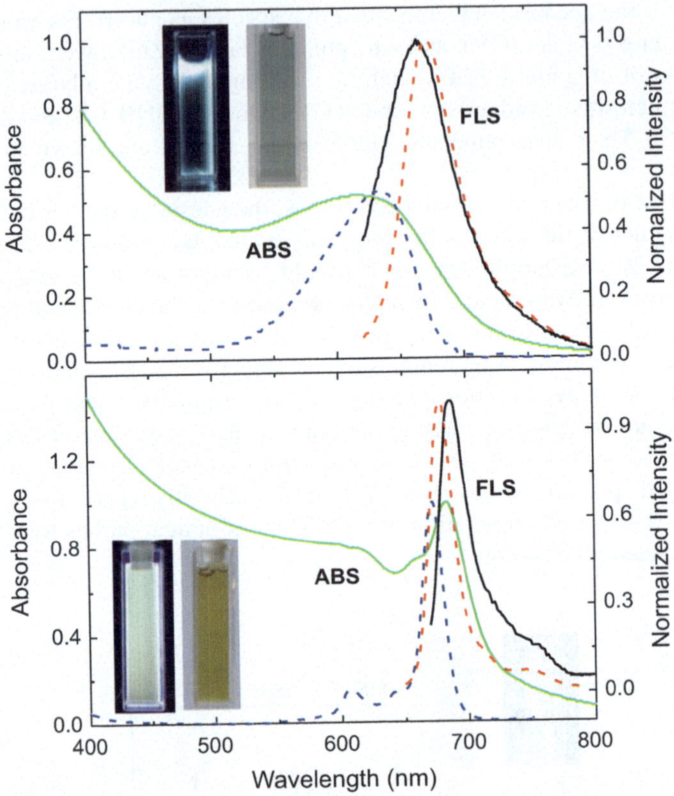

Fig. 7.3 Absorption (ABS) and fluorescence (FLSC) spectra of NB@CDots (upper, —) and ZnPc@CDots (lower, —) and the corresponding free dyes (- - -) in aqueous solutions (except for free ZnPc in DMSO). Insets: Photographs of aqueous solutions of the corresponding samples under UV in the dark (left) and under natural day light (right). (Originally Figure 3 in Sun et al. (2015). Reproduced with permission)

1.2 Unique Photoexcited State and Redox Properties

CDots are known and thus widely explored for their unique photoinduced redox characteristics. Interestingly and surprisingly, even more unique redox behaviors were found in photoexcited G@CDots (LeCroy et al. 2019), with major implications both mechanistically and technologically. The surprising finding for the G@CDots (e.g., G = NB) was that the effect associated with the encapsulation/embedding of the dye species in the dot core could go much beyond the originally anticipated host–guest relationship. "More specifically, the CDots with dye modification would adopt most of the dye's absorption and fluorescence emission features, yet at the same time retain the unique photoinduced redox characteristics intrinsic to non-modified neat CDots as both excellent electron acceptors and donors. The results suggest that the specific hybrid CDots platform based on the modification of core carbon nanoparticles by dyes or other chromophoric species may ultimately enable the preparation of CDots with desired optical and redox properties in a by-design fashion" (LeCroy et al. 2019).

As reported (LeCroy et al. 2019), "optical absorption spectra of the PEG-NB@CDots (NB = nile blue as guest, and PEG moieties on the dot surface) are obviously different from that of the PEG-CDots without the NB modification (Fig. 7.4), and in the former the NB-derived species must be responsible for the extra absorptions at longer wavelengths. When excited into those wavelengths, 530 nm for example, bright red fluorescence emissions from the PEG-NB@CDots were observed (Fig. 7.4). The quantum yields for the red emissions were found to be very high, 85–90% in a methanol solution determined by cross-referencing to three fluorescence standards, one of which was free NB in methanol with a known fluorescence quantum yield of 27%. The corresponding fluorescence lifetime of the PEG-NB@CDots was also found to be long, 3.6 ns vs 1.25 ns in free NB. Such an ultrahigh fluorescence emission performance is beyond a simple explanation such that the NB-derived species become more fluorescent when confined in the more restrictive core carbon nanoparticle structure. More likely, in the modified CDots the carbon nanoparticles and the modifying dye species are not only physically incorporated, which may be considered structurally as host and guest, respectively, but also electronically interacted in such a strong fashion to result in altered or new electronic states and transitions shared by the host and guest. Thus, electronically the host-guest pair may have to be considered as one electronically integrated system instead of two segregated entities merely being together physically in a simple host-guest arrangement. Consistent with such a mechanistic picture are the negative outcomes highlighted below from experiments aimed at probing the possible presence of intra-dot host-to-guest excited state energy transfers."

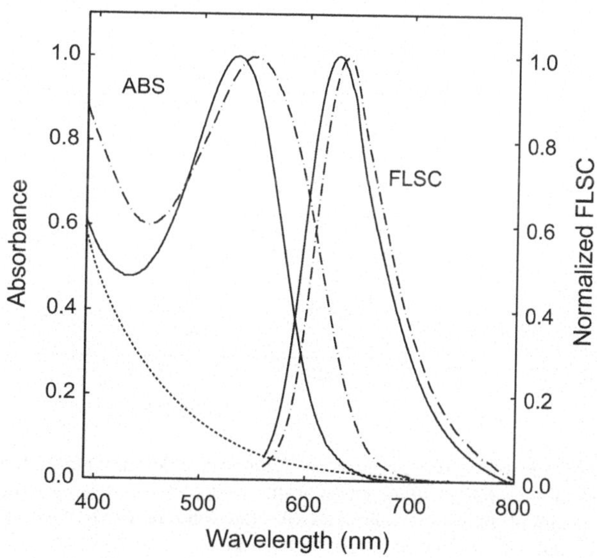

Fig. 7.4 Absorption (ABS) and fluorescence (FLSC, 530 nm excitation) spectra of the PEG-NB@CDots in methanol (solid) and aqueous (dash-dot) solutions and for comparison ABS of the PEG-CDots in methanol (dot). (Originally Figure 2 in LeCroy et al. (2019). Reproduced with permission)

For the red emission band (centered around 630 nm, Fig. 7.4), observed quantum yields shown in Fig. 7.5 "suggest no significant energy transfers from the photoexcited carbon nanoparticles (the host, with dominating absorptions in the blue/green) to the emissive excited states of the modifying NB-derived species (the guest), because otherwise the observed quantum yields at blue/green excitations would be much higher to be comparable with those at red excitations, which is obviously not the case as shown in Fig. 7.5. The fluorescence lifetime results (Fig. 7.5) are consistent with the assessment based on the quantum yields. The energy transfer would have to be expected if the two entities in each core carbon nanoparticle, namely the nanoscale carbon host and the guest dye species, were only physically in contact or mixed while electronically largely independent. Thus, the observed lack of excited state energy transfers suggests that the two entities must be more integrated electronically" (LeCroy et al. 2019). In fact, in a different configuration of CDots with surface-tethered organic dyes, substantial fluorescence resonance energy transfers (FRET) were reported (Huang et al. 2013), namely, in that configuration CDots as donor and dyes as acceptor for the excited state energies must be two electronically independent entities.

Strongly supporting the G@CDots being "one electronically integrated entity" were the results on their photoinduced redox characteristics. When excited into the absorption region dominated by the NB-derived species (e.g., 530 nm) in PEG-NB@CDots, the red emission quantum yields and lifetimes could be quenched efficiently

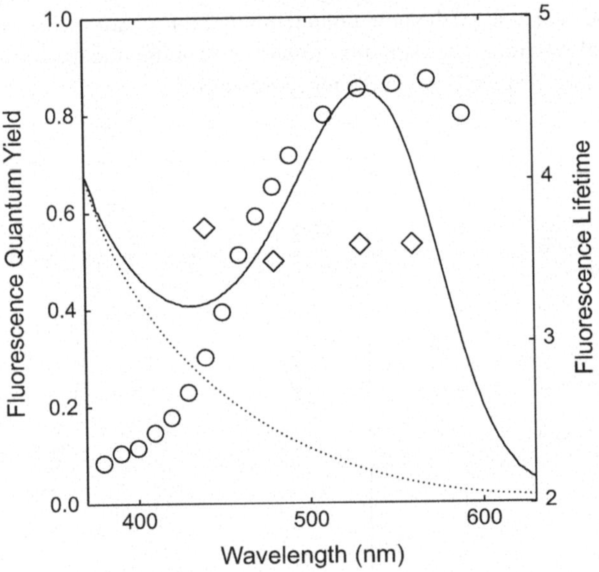

Fig. 7.5 Fluorescence quantum yields (circle, left axis) and lifetimes (diamond, right axis) of the PEG-NB@CDots in methanol at different excitation wavelengths and for referencing absorption spectra of the sample solution (solid line) and the PEG-CDots in methanol (dot line). (Originally Figure 3 in LeCroy et al. (2019). Reproduced with permission)

(at the upper limit of diffusion control) by both electron acceptor 2,4-dinitrotoluene (DNT) and donor *N,N*-diethylaniline (DEA) (LeCroy et al. 2019). For comparison, fluorescence emissions of free NB could be quenched by DEA, but not DNT. "Thus, even with the photon-harvesting due almost exclusively to the NB-derived species at the longer wavelengths (Fig. 7.4), the photoinduced redox characteristics of the PEG-NB@CDots is still clearly dictated by what is intrinsic to carbon nanoparticles without any modifications, the same as that in non-modified neat CDots. Again, the PEG-NB@CDots obviously behave individually as a single electronically integrated entity in the response to or interactions with external electron acceptors and donors" (LeCroy et al. 2019).

A simple translation for the observed surprising redox behavior is that in the G@CDots the optical absorption for photon-harvesting could be enhanced substantially over the longer wavelength spectral region by the guest dye species, but the photon energies thus harvested would not be localized in the dye species, rather shared by the G@CDots as a whole. Obviously this can be extremely valuable to exploiting the photoinduced redox properties of CDots for optoelectronics and related applications. The results discussed above "show that the core carbon nanoparticles could be modified in the nanoparticle structure by a selected chromophore to obtain the specific type of hybrid CDots that are not only of the targeted optical absorption profiles for considerably more effective photon-harvesting but also with the redox characteristics intrinsic to CDots preserved and enhanced." "Since there are a wide variety of selections among the chromophoric species for the modification purpose, this type of hybrid CDots represents a highly versatile and effective platform for the design and preparation of zero-dimensional carbon nanostructures with specifically targeted optical, redox, and other properties" (LeCroy et al. 2019).

1.3 Red/Near-IR Fluorescence Bioimaging

A major motivation in the development of host–guest CDots with red/near-IR dyes or dye-derived species as the guest is for bioimaging in the spectral window more transparent to tissues and biological species in general. The G@CDots (G = NB or CV, Figs. 7.3 or 7.2, respectively) have already been demonstrated as bright fluorescence probes for bioimaging in vitro and in vivo (Wang et al. 2017; Yang et al. 2018a).

As reported (Wang et al. 2017), the NB@CDots were used as probes in the fluorescence imaging of the Sprague-Dawle (SD) rat mesenchymal stem cells (MSCs). The results suggested that the NB@CDots in the live MSCs were mostly in the cell cytoplasm, not in the nucleus (Fig. 7.6), no different from the cellular distributions of conventional CDots without guests in general. The live MSCs labeled with the NB@CDots could be stained by the established lysosome staining agent LysoSensort Green DND-189 for their being co-localized in the fluorescence imaging, with observations in the red and green emission windows, respectively. In Fig. 7.7, the yellow spots in the merged images were assigned to the overlapping of the red and green signals, namely, the lysosomal co-localization of the NB@CDots with the

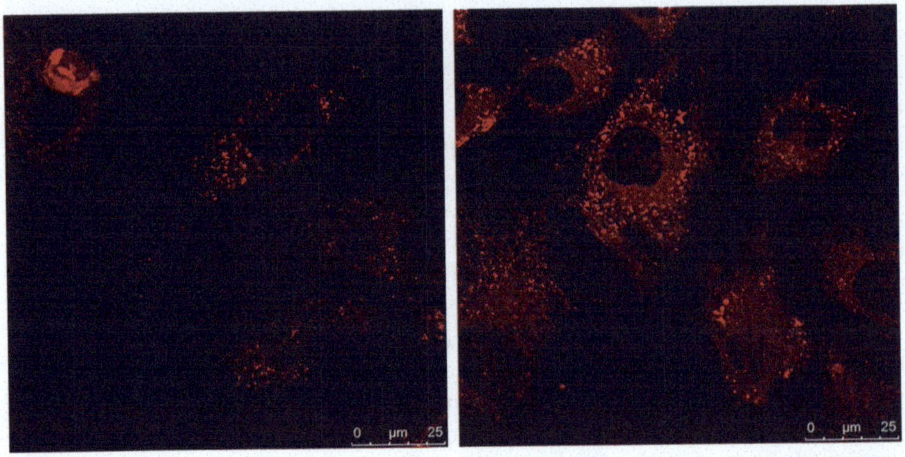

Fig. 7.6 Confocal fluorescence images (excitation at 488 nm, and emissions monitored at 608–720 nm) for live MSCs labeled with the NB@CDots at dot concentrations of 0.25 μM (left) and 0.5 μM (right). (Originally Figure 8 in Wang et al. (2017). Reproduced with permission)

specific lysosome staining agent. There were also intracellular areas with red signals only, suggesting that some of the NB@CDots could escape from the endosomes into the cytoplasm, a phenomenon also found in similar cell imaging with other nanoparticles.

It was concluded that for the fluorescence labeling of stem cells, the host–guest CDots could apparently not only be taken up by cells like their neat counterparts without guests but also offer much enhanced performance in the red/near-IR, more desirable in terms of suppressing background signals in cell imaging and eventually for the labeled cells to be used for in vivo tracking and other relevant applications (Wang et al. 2017).

The CV@CDots (Fig. 7.2) were used as red/near-IR fluorescence probes for imaging in mice (Yang et al. 2018a). Upon subcutaneous injection of the dots into normal nude mice, bright red fluorescence on the back of the mouse was observed (Fig. 7.8), which with the efficient spectral unmixing could be distinguished from that of autofluorescence (tinted as green in the figure). As determined, the contrast between fluorescence signals from the dots and the background was high, suggesting that the CV@CDots could indeed serve as excellent probes for fluorescence imaging in vivo.

Nude mice bearing tumor (papillary thyroid carcinoma cell line TPC-1) were also used for intratumor injection and the imaging evaluation in vivo (Yang et al. 2018a). Strong red fluorescence emissions were observed from the tumor site immediately after the injection (Fig. 7.9). The red emissions were associated with the tumor only, without any significant diffusion to the nearby tissue. The signal decayed slowly over time, with the emission intensity at the tumor site decreased to about one-third of the initial intensity after 9 h (Fig. 7.9), but still no meaningful signal found in other tissues around the tumor.

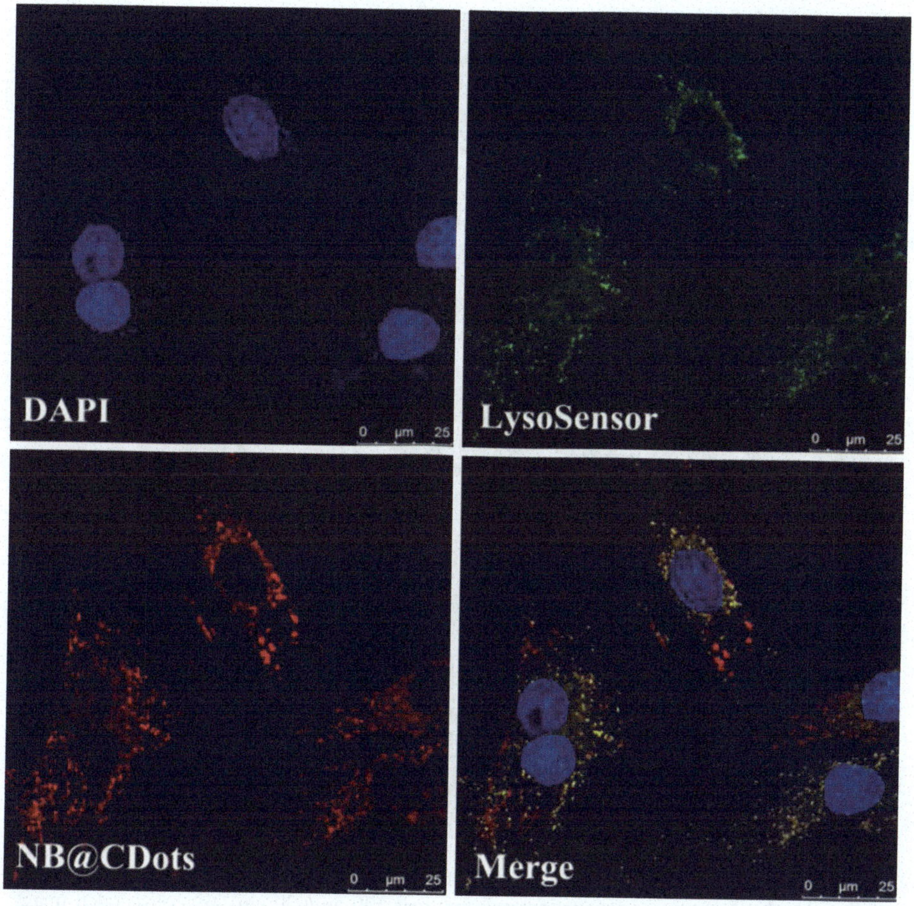

Fig. 7.7 Confocal fluorescence images of live MSCs with NB@CDots, LysoSensor™ Green DND-189, and DAPI obtained under different conditions. Upper-left: excitation at 405 nm for DAPI and emissions monitored at 424–469 nm; upper-right: excitation at 488 nm for LysoSensor™ Green DND-189 and emissions monitored at 490–539 nm; lower-left: excitation at 488 nm for NB@CDots and emissions monitored at 608–720 nm; and lower-right: the merged three images. (Originally Figure 9 in Wang et al. (2017). Reproduced with permission)

For the intravenous injection and then circulation for 30 min, there were red fluorescence emissions from the probes in the abdomen area (Fig. 7.10) with the most intense signal from the bladder and significant but much weak signal from the small intestine, suggesting that the dots could be excreted through both urine and bile. Of particular interest was the observation of significantly stronger signals from the tumor (Fig. 7.10). The tumor tissue was much more fluorescent than the nearby skin tissue, even though the total fluorescence intensity of tumor was still lower than that of bladder. The abundant blood vessels in tumor must have facilitated the accumulation of the fluorescence probes. The relatively significant fluorescence signals from the ear, paws, and tail could be attributed to the thinner skin in those areas.

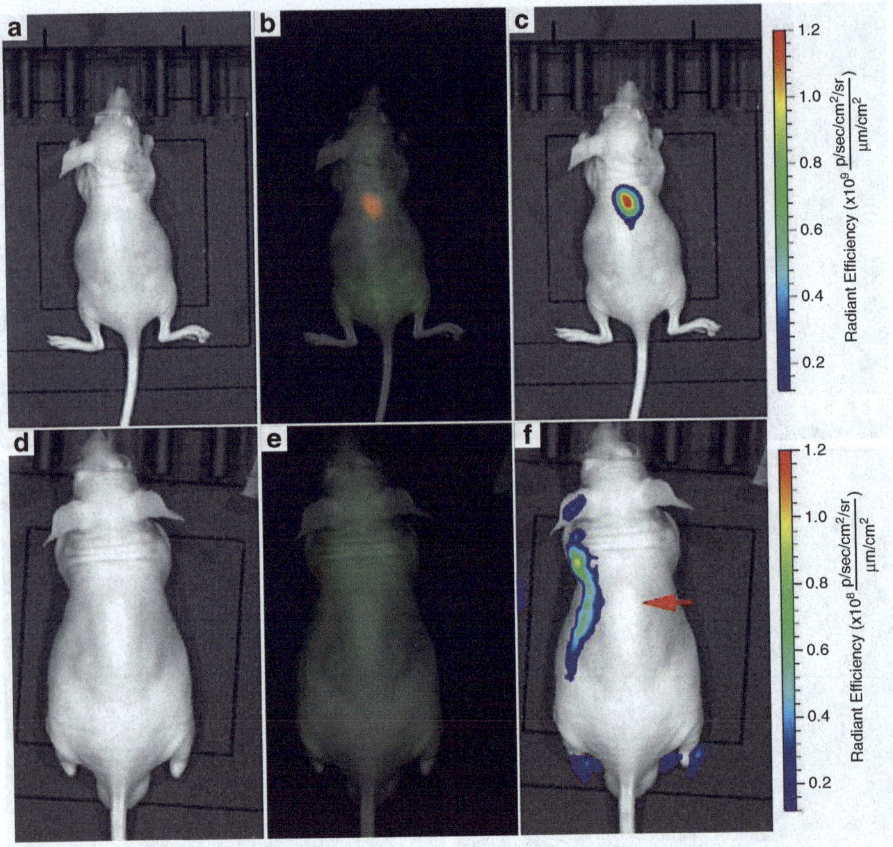

Fig. 7.8 Images with subcutaneous injection of the PEG-CV@CDots at 10 min post-injection, (**a**) bright field, (**b**) fluorescence (excitation/emission pair 520 nm/620 nm), (**c**) color-coded image, and at 24 h post-injection, (**d**) bright field, (**e**) fluorescence, (**f**) color-coded image (the injection area indicated by the arrow). (Originally Figure 4 in Yang et al. (2018a). Reproduced with permission)

At 6 h after the intravenous injection, the mice were sacrificed to collect the organs for ex vivo imaging (Yang et al. 2018a). The results collectively suggested again the fast and nearly complete excretion of the CV@CDots from mice, rather similar to what were observed in the in vivo imaging with neat CDots without guest dyes (Yang et al. 2009a; Huang et al. 2013), also providing further evidence for the nontoxic nature of CDots in vivo. It was concluded on the G@CDots platform that "these red/near-IR CDots are indeed excellent fluorescence probes for bioimaging *in vivo*, including also their apparently enhanced tumor uptake. With their further development and validation, broad applications of these high-performance and non-toxic probes in bioimaging and theranostics in general may be envisaged" (Yang et al. 2018a).

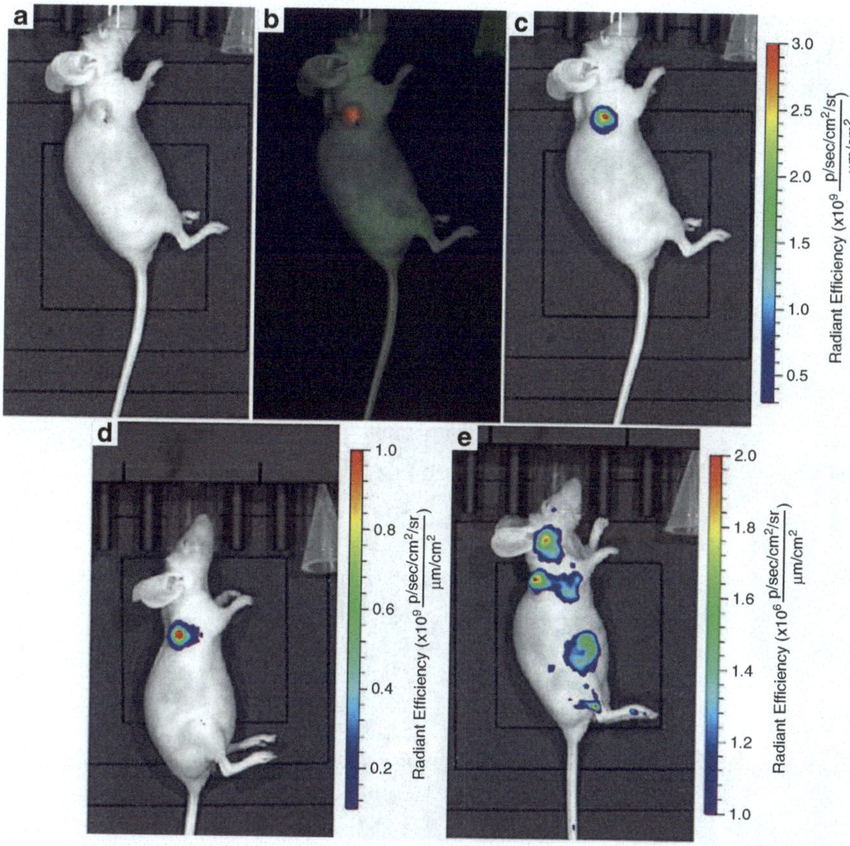

Fig. 7.9 Images with intratumor injection of the PEG-CV@CDots. At 10 min post-injection: (**a**) bright field; (**b**) fluorescence; and (**c**) color-coded. (**d**) and (**e**) are color-coded images at 9 h and 48 h post-injection, respectively. (Originally Figure 5 in Yang et al. (2018a). Reproduced with permission)

2 Carbon/Metal Oxide Hybrid Dots

Carbon is apparently compatible with many metals and metal oxides (including some famous semiconductors) in terms of the carbon nanoparticles being surface-doped/surface-coated by, structurally encapsulating, or in composition with the metals/metal oxides. These carbon-based/carbon-derived nanostructures, generally referred to as modified carbon nanoparticles, can be surface passivated to obtain various hybrid CDots.

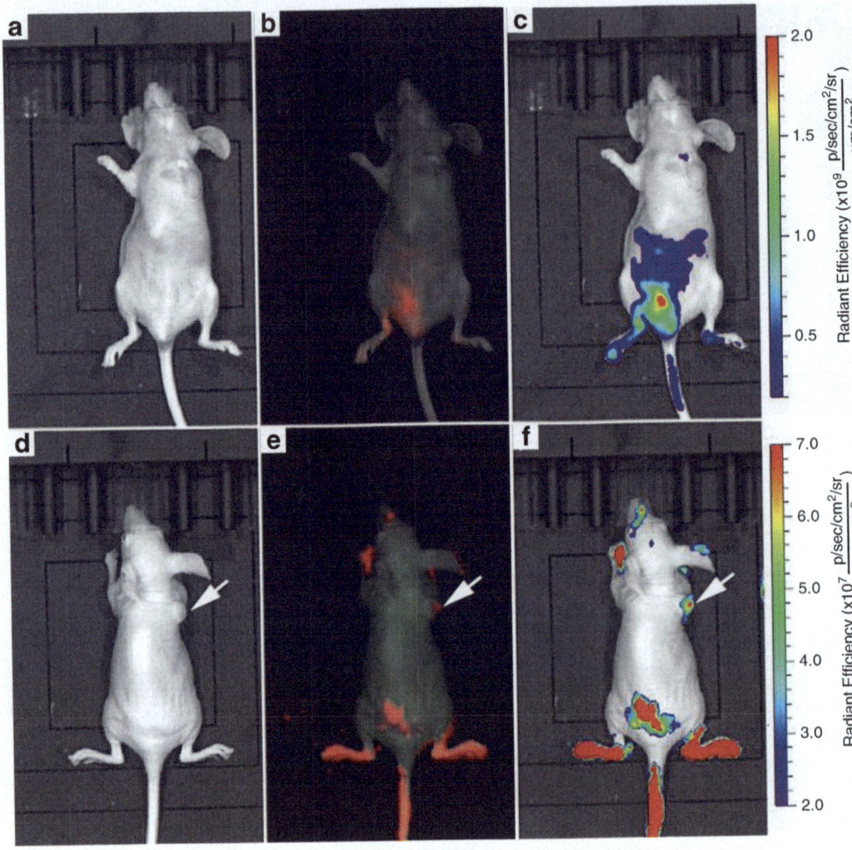

Fig. 7.10 Images with intravenous injection of the PEG-CV@CDots at 30 min post-injection. Face up: (**a**) bright field; (**b**) fluorescence; and (**c**) color-coded. Face down: (**d**) bright field; (**e**) fluorescence; and (**f**) color-coded. The tumor was indicated by the white arrows. (Originally Figure 6 in Yang et al. (2018a). Reproduced with permission)

2.1 *CDots from Surface-Doped Carbon Nanoparticles*

Historically in the development of conventional semiconductor QDs, CdSe nanocrystals as a representative example, the surface capping of the nanocrystals by wide-bandgap semiconductors like ZnS was thought to be responsible for the much enhanced fluorescence brightness in the resulting CdSe/ZnS core-shell nanoparticles. The enhancement was rationalized as being due to the surface capping by the wide-bandgap semiconductor reducing or preventing the unwanted nonradiative deactivation of the CdSe photoexcited states. From the very beginning, CDots were considered as being mechanistically different from semiconductor QDs. Nevertheless, the small carbon nanoparticles were still doped/coated with nanoscale wide-bandgap semiconductors as an exploration, in which surprisingly strong effects on the optical properties of the corresponding CDots were observed (Sun et al. 2008).

Experimentally, the aqueous dispersed small carbon nanoparticles harvested in the typical processing and selection protocol were surface-doped with ZnO or ZnS. Here the wording of "doped" instead of "coated" was for the reason that the nanoparticles without any exposed carbon surface, namely, with a more complete ZnO or ZnS coating on the surface, were excluded in the experimental procedures. The ZnO- or ZnS-doped carbon nanoparticles were functionalized by PEG_{1500N} (Fig. 4.1) to yield CDots of ZnO- or ZnS-doped carbon cores, denoted as C_{ZnO}-Dots or C_{ZnS}-Dots, respectively (Figs. 7.11 and 7.12) (Sun et al. 2008).

A pleasant surprise was the finding that C_{ZnO}-Dots and C_{ZnS}-Dots were much more fluorescent than their corresponding non-doped CDots. The results were rationalized with a proposal "that the dopant may provide secondary yet more effective surface passivation in combination with the organic passivation agents. Results from repeated control experiments suggested that the functionalization of ZnO- or ZnS-doped carbon nanoparticles by the organic (PEG_{1500N}) molecules is necessary for the observed very strong photoluminescence" (Sun et al. 2008).

Later, CDots with TiO_2-doped carbon nanoparticle cores, thus C_{TiO2}-Dots, were added to the list (Anilkumar et al. 2011), for which the motivation was related to the extreme popularity of nanoscale TiO_2. Also in that work (Anilkumar et al. 2011), as-prepared samples of C_{ZnS}-Dots and C_{TiO2}-Dots were fractionated on an aqueous gel column to harvest more fluorescent fractions. In fact, the C_{ZnS}-Dots and C_{TiO2}-Dots in the best fractions thus separated, with quantum yields of 78% and 70%, respectively (Anilkumar et al. 2011), still hold the record to this day as the most fluorescent CDots ever made with pre-existing carbon nanoparticles.

For mechanistic elucidation, fluorescence decays of C_{TiO2}-Dots and their corresponding non-doped CDots were compared and found to be essentially the same (Fig. 7.13). Thus, it was argued that in the mechanistic framework for CDots (Fig. 4.7), the doping effect must be enhancing primarily the formation of the emissive excited states, namely, the equivalent of "negative static quenching" (compared to

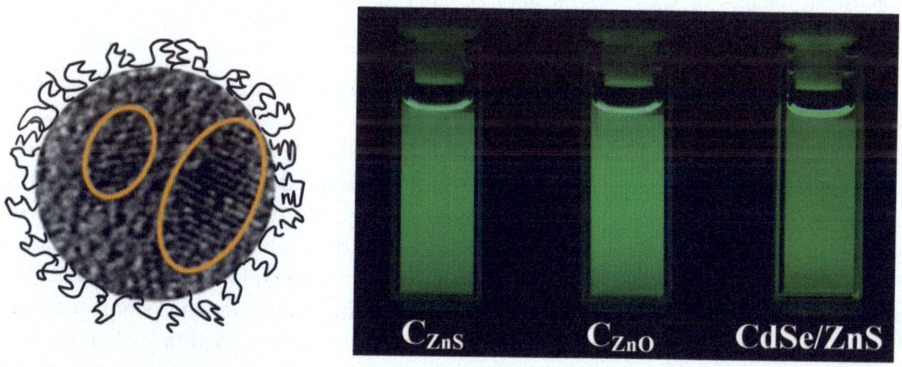

Fig. 7.11 Left: A cartoon illustration on CDots with a doped carbon core (from an experimental HR-TEM image with ZnS lattice fringes circled); right: aqueous solutions of C_{ZnS}-Dots and C_{ZnO}-Dots (450 nm excitation for both) compared with commercial toluene solution of CdSe/ZnS dots (matching optical density at excitation), all photographed through a 475 nm cutoff filter. (Originally Figure 1 in Sun et al. (2008). Reproduced with permission)

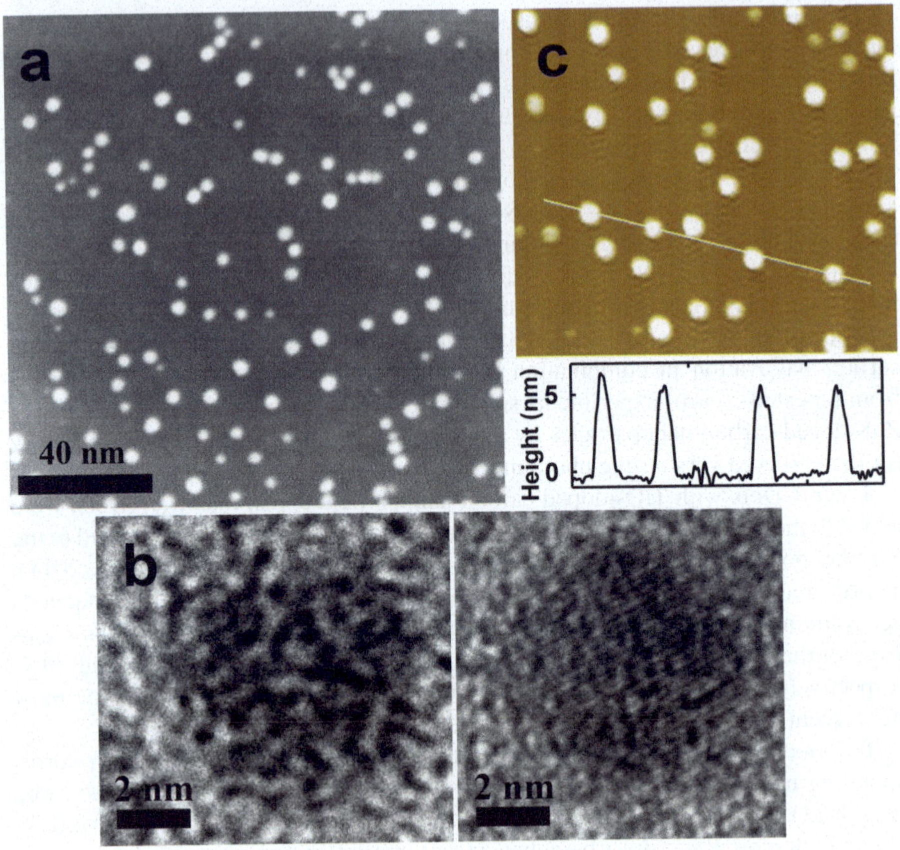

Fig. 7.12 (**a**) TEM (Z-contrast) images of C_{ZnS}-Dots. (**b**) High-resolution TEM images of individual CDots without doping (left) and with ZnS-doping (right, showing lattice fringes). (**c**) AFM topography images of C_{ZnO}-Dots on mica substrate (and the height profile along the line). (Originally Figure 2 in Sun et al. (2008). Reproduced with permission)

the positive static quenching observed with electron donor DEA and electron acceptor DNT as quenchers discussed in Chap. 6).

The following edited version of the description in (Anilkumar et al. 2011) may be valuable to the further investigation and understanding of these CDots with doped carbon nanoparticle cores. "Mechanistically, the fluorescence emissions in CDots are attributed to radiative recombinations of the carbon nanoparticle surface-trapped electrons and holes, where the large surface area (relative to the nanoparticle volume) and diverse surface energy trapping sites in the small carbon nanoparticles are stabilized by the surface passivation agents. Within such a mechanistic framework, the enhanced fluorescence performance in nanoscale semiconductor-doped CDots, more specifically C_{ZnS}-Dots and C_{TiO2}-Dots, may be rationalized in terms of improved surface passivation by a combination of inorganic surface-doping and organic functionalization, as well as probably more diverse surface sites as a result of the doping (thus facilitating more effective trapping of elec-

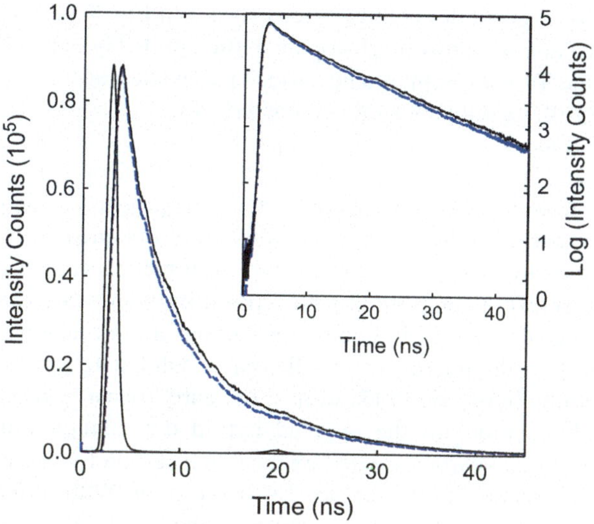

Fig. 7.13 Fluorescence decays (407 nm laser excitation, and monitored through 470 nm bandpass filter) of the most fluorescent C_{TiO2}-Dots (—) and the highly fluorescent non-coated CDots (- - -). (Originally Figure 3 in Anilkumar et al. (2011). Reproduced with permission)

trons and holes on the nanoparticle surface)." Among several mechanistic and technical issues for consideration, one is on the selection of the dopant, for which nanoscale semiconductors are advantageous but apparently not necessary. In fact, in the evaluation on using other nanomaterials such as nanoscale AgCl for the doping (Wang 2010), there was also enhancement in the observed fluorescence brightness of the doped CDots, thus providing evidence for the notion that the surface doping effect might have little to do with whether or not the selected nanomaterials for the doping are semiconductors. "Structural and/or material parameters such as the compatibility of the selected inorganic salt with the carbon nanoparticle surface may play major roles in determining the fluorescence performance of the resulting CDots. Another issue is on the relative quantities of the inorganic and organic surface passivation moieties, or perhaps even their mutual compatibility on the carbon nanoparticle surface" (Anilkumar et al. 2011).

C_{TiO2}-Dots represent just one of many possible and practical carbon-TiO_2 hybrid dot configurations, but due to their superior fluorescence properties, they may serve as the benchmark in the study of other TiO_2-containing hybrid CDots.

2.2 *Carbon/TiO₂ Hybrid Dots*

In carbon/TiO_2 hybrid dots, there are various structural arrangements, each of which can have different contents of nanoscale TiO_2 in different sizes. Some of these variations have resulted in dramatically different properties in the corresponding hybrid dots. Among the selected dot configurations shown in Fig. 7.14, C_{TiO2}-Dots apparently represent a unique platform, clearly an outlier, probably due to the fact that the role of nanoscale TiO_2 as an inorganic surface passivation agent in the hybrid dots has little to do with its semiconductor characteristics.

The hybrid dots of a small TiO_2 nanoparticle as the core and a carbon shell surface-passivated by organic species, denoted as C/TiO_2-Dots (Fig. 7.14), are obviously different from C_{TiO2}-Dots in photoexcited state properties and processes (Liu et al. 2018). Experimentally, the C/TiO_2-Dots were obtained from a protocol in which the pre-prepared and selected small TiO_2 nanoparticles (average sizes around 8 nm in diameter) were composited with (coated by) nanoscale carbon and functionalized by organic species in controlled carbonization reactions. The reaction conditions based on the microwave irradiation of solid-state mixtures of the TiO_2 nanoparticles with oligomeric PEGs were sufficiently robust to ensure the required substantial carbonization for the targeted hybrid dot configuration (Fig. 7.14), namely, each TiO_2 nanoparticle coated with enough carbon for essentially a carbon shell with PEG moieties on the surface for passivation. With different PEG/TiO_2 ratios and more importantly microwave processing conditions, two C/TiO_2-Dots samples with TiO_2 contents of 31 wt% (6.3 mol%) and 71 wt% (27 mol%) in the core nanoparticles of the dots (excluding carbons in the surface PEG moieties) were

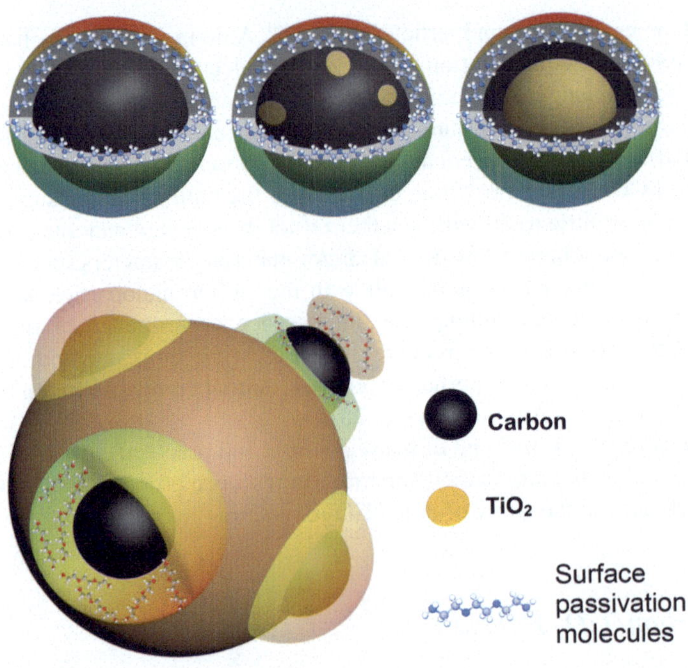

Fig. 7.14 Cartoon illustrations on a neat CDot (upper-left) and carbon-TiO_2 hybrid dots in different configurations, including those with a TiO_2-doped carbon nanoparticle core (upper-middle), a small TiO_2 nanoparticle with a carbon shell (upper-right), and a large TiO_2 nanoparticle surface decorated with nanoscale carbon domains (lower), all with surface passivation by organic species

prepared. As a reference, CDots without TiO_2, referred to as PEG-CDots, were prepared in the same microwave processing (Liu et al. 2018).

As reported (Liu et al. 2018), "Optical absorption spectra of the C/TiO_2-Dots (Fig. 7.15) in the visible spectral region must be due to the carbon domains because of the absorption of small carbon nanoparticles (Fig. 7.15). Between the two C/TiO_2-Dots samples, while their absorption spectral profiles were essentially the same, their absorbances were found to be proportionally different with respect to the carbon contents in the cores of the C/TiO_2-Dots." On their fluorescence emissions, both spectra similar to that of PEG-CDots (Fig. 7.15), "these carbon-TiO_2 hybrid dots are obviously less fluorescent than their neat counterpart, with observed quantum yields of about 1% for the C/TiO_2-Dots of 31 wt% TiO_2 and about 0.6% for the C/TiO_2-Dots of 71 wt% TiO_2 versus the yield of 3% for the PEG-CDots. The lower quantum yields in both C/TiO_2-Dots samples and the relatively even lower in the sample with more TiO_2 suggest the likely role of TiO_2 in the hybrid dots as quencher for the emissive excited states of the carbon domains. In fact, the quenching should be expected because configuration-wise the C/TiO_2-Dots are equivalent to dye-sensitized semiconductors (TiO_2), with the carbon domains serving the dye function for photon-harvesting in the visible spectral region. What might be considered surprising is that the fluorescence emissions are not diminished even with the TiO_2 being in such close contact with the carbon domains in the hybrid dots. One explanation may be that the carbon domains are substantial in size, not a thin coating on the surface of the TiO_2 nanoparticles" (Liu et al. 2018).

For the hybrid dots of a similar configuration but based on much larger TiO_2 nanoparticles of 25 nm in diameter, denoted as C/$TiO_2^{(25nm)}$-Dots (Figs. 7.14 and 7.16), the same microwave-assisted thermal carbonization under more vigorous processing conditions was used in the synthesis (Pan et al. 2020). A major structural

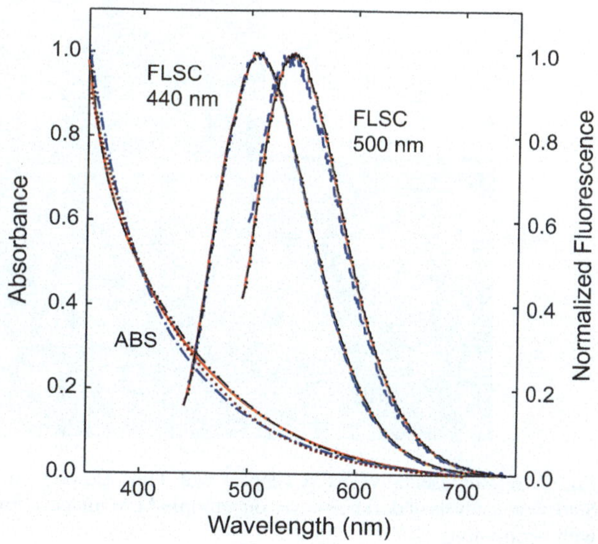

Fig. 7.15 Absorption (ABS) and fluorescence (FLSC, with the excitation wavelengths shown) spectra of the C/TiO_2-Dots with 71 wt% (——) and 31 wt% (-..-..-) TiO_2 and the PEG-CDots (-.-.-). The absorption spectrum of the aqueous suspended small carbon nanoparticles (•••••••) is also shown for comparison. (Originally Figure 7 in Liu et al. (2018). Reproduced with permission)

difference with the much larger TiO_2 nanoparticles in the hybrid dots was that the nanoscale carbon domains were insufficient to form a carbon shell, leaving some exposed TiO_2 surface (Fig. 7.14), which apparently resulted in significantly different properties.

It was found that optical absorption and fluorescence emission spectra of C/$TiO_2^{(25nm)}$-Dots were similar to those of C/$TiO_2^{(8nm)}$-Dots (with TiO_2 nanoparticles of about 8 nm in diameter in the same hybrid dot samples discussed above), but the observed fluorescence quantum yields were less than 1% (Pan et al. 2020). The major differences in photoexcited state properties between C/$TiO_2^{(25nm)}$-Dots and C/$TiO_2^{(8nm)}$-Dots were reflected more clearly by their dramatically different photodynamic effects. Such effects were credited in earlier studies for the observed antibacterial activities of CDots under visible light illumination (Meziani et al. 2016; Dong

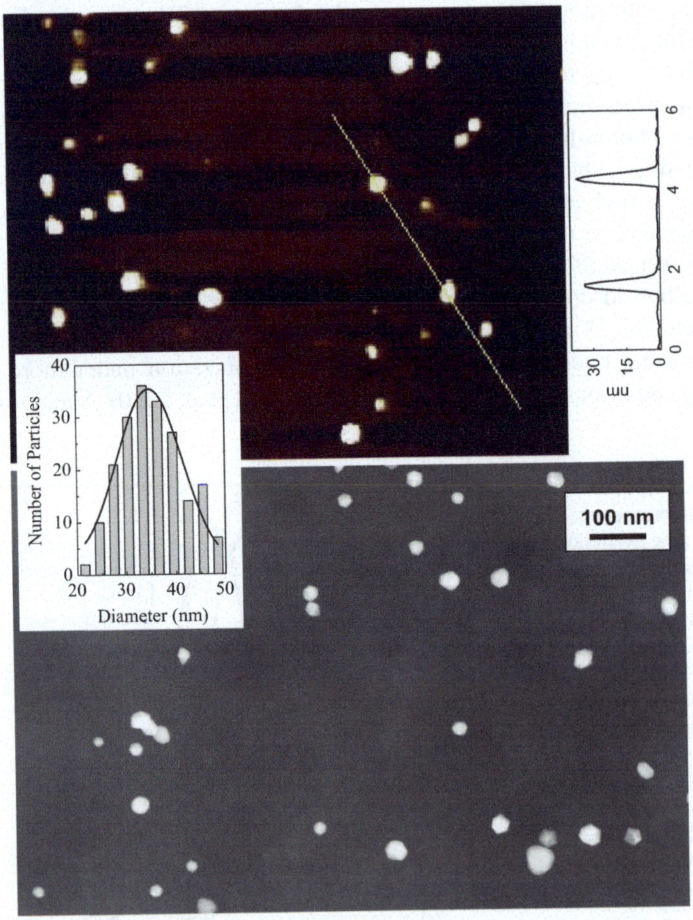

Fig. 7.16 Representative AFM (upper) and TEM (lower) images of C/$TiO_2^{(25nm)}$-Dots. Inset: Statistical analysis of dot sizes based on multiple AFM images. (From Pan et al. (2020). Reproduced with permission)

et al. 2020), as discussed in Chap. 6. For the C/$TiO_2^{(8nm)}$-Dots against the laboratory model bacteria *Bacillus subtilis* under visible light, the inhibition effects were limited, much weaker than those with the use of C/$TiO_2^{(25nm)}$-Dots (Pan et al. 2020).

For the C/$TiO_2^{(25nm)}$-Dots treating *B. subtilis* in the same experimental protocol at dot concentrations (in terms of carbon-equivalent concentrations for the nanoscale carbon domains in the cores of the hybrid dots, Fig. 7.14) up to 0.1 mg/mL under visible light for 2 h, the viable cell number was reduced by more than 6 logs (Fig. 7.17).

The observation of substantially more effective antibacterial actions by the hybrid dots with the larger TiO_2 nanoparticles was surprising for several reasons. Because of the larger TiO_2 nanoparticles and therefore much larger dot sizes, the light scattering must be more significant in their associated solutions, which would interfere with the light absorption in the photo-irradiation experiments. The interference might be more pronounced for the specific dot configuration, with the nanoscale carbon domains responsible for the light absorption being right next to the larger TiO_2 nanoparticle. The overall larger dot sizes would also be unfavorable in comparison with their smaller counterparts in interactions with the targeted bacterial cells. Therefore, there must be a mechanistic explanation for the surprising outcome. As a rationale (Pan et al. 2020): "In all these C/TiO_2 hybrid dots, the TiO_2 nanoparticles are effective 'quenchers' for the excited states produced by the light absorption of the nanoscale carbon domains, as reflected by the observed substan-

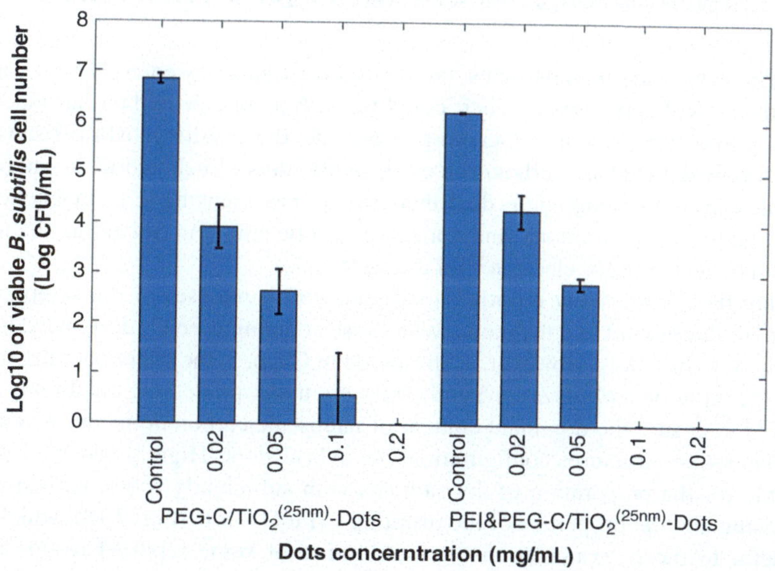

Fig. 7.17 Logarithmic viable cell numbers of *B. subtilis* in untreated control samples and the samples treated with two C/$TiO_2^{(25nm)}$-Dots samples (one with PEG on the surface and the other with a mixture of PEI and PEG on the surface) at concentrations ranging from 0.02 mg/mL to 0.2 mg/mL for 2 h under illumination of a 60 W-equivalent daylight LED. (From Pan et al. (2020). Reproduced with permission)

tially weaker fluorescence emissions. This is analogous to what is known in dye-sensitized TiO_2 systems, namely the nanoscale carbon domains essentially serve the function of 'dye-sensitizers.' As a result of the quenching effect, the antibacterial activities by the surface-passivated carbon domains are substantially 'quenched' as well. Thus, the observed bacteria-killing actions by all these C/TiO_2 hybrid dots must be contributed majorly or dominated by the sensitized TiO_2 nanoparticles. In such a mechanistic framework, the small TiO_2 nanoparticles are more enclosed by the nanoscale carbon domains than their larger sized counterparts in the corresponding hybrid dots, which may hinder their killing action to result in less effective viable cell reductions" (Pan et al. 2020).

There are mechanistically significant and also technically valuable implications of the comparisons between different carbon–TiO_2 hybrid dot configurations discussed above. With the variable overall dot sizes, structural arrangements, and compositions building on nanoscale TiO_2 domains of different dimensions, the hybrid dots may be designed and prepared for many specific applications. Moreover on the hybrid dots based on the large TiO_2 nanoparticles, these nanoparticles are commercially available and very inexpensive, so their derived hybrid dots may find a broad range of uses as effective visible/natural light-activated antimicrobial agents and in other technological fields.

2.3 *Carbon-Based/Carbon-Derived Magnetic Hybrid Dots*

Magnetic iron oxide nanoparticles have been investigated extensively for a variety of technological applications. Their coupling with nanoscale carbon introduces an optical component, among other properties, into the resulting hybrid nanostructures, simply denoted as "carbon/iron oxide hybrid dots." Such hybrid dots may find valuable uses in bioimaging (as dual-modality probes), sensing (e.g., magnetic separation and/or concentration), and optical–magnetic coupling (including the use as a materials platform for electron spin interactions).

There have been many reports on carbonization syntheses of dot samples that were both fluorescent and magnetic, with those containing Fe_3O_4 designated as carbon/Fe_3O_4 hybrid dots. However, as discussed in Chap. 5, the overwhelming majority of the reported carbonization syntheses were under processing conditions much too mild to ensure the dominant content of nanoscale carbon domains, as required for CDots-like structures/configurations in hybrid dots. Highlighted here are an example for the preparation of dot samples with sufficiently robust carbonization processing and the properties of the resulting C/Fe_3O_4-Dots (Fig. 7.18), which may be useful to the re-examination or recalibration of some reported results in the literature.

In the investigation reported in (Hu et al. 2017), "the PEGylated hybrid dots PEG-C/Fe_3O_4-Dots were prepared in the one-pot thermal carbonization processing of precursor mixtures containing oligomeric PEGs (average molecular weight ~ 1500, PEG_{1500}) and ferric ammonium citrate (FAC) in various PEG/FAC mass ratios.

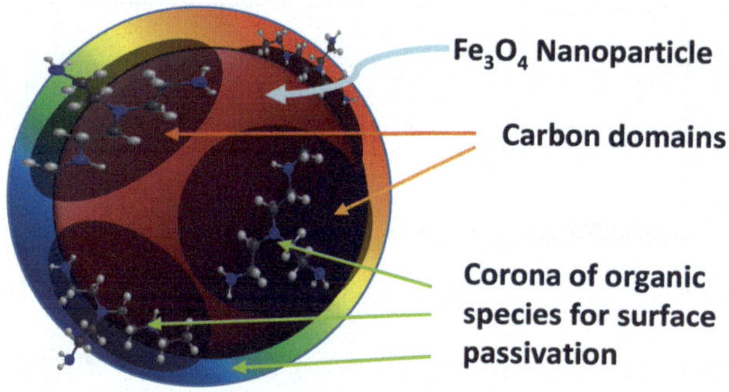

Fig. 7.18 A cartoon illustration on a carbon/Fe_3O_4 hybrid dot with organic species for the desired surface passivation effect. (Originally Figure 1 in Hu et al. (2017). Reproduced with permission)

Different processing conditions involving the use of microwave irradiation for the thermal carbonization were evaluated, including various combinations of microwave power and irradiation time duration in a single treatment and the use of multiple treatment cycles, each of which consisted of microwave heating and then a quick cooling in the ambient. The reaction mixture from each processing was dispersed in water, and the resulting aqueous dispersion was separated in terms of magnetic precipitation in a magnetic separator with a permanent magnet to harvest the magnetically responsive fraction. The evaluation of the outcomes associated with different PEG/FAC mass ratios and thermal carbonization conditions was based on parameters including the product yield for the dot sample, the aqueous dispersibility, and the C/Fe ratio in the sample from the magnetic separation. The use of a PEG/FAC ratio of 2 in the precursor mixture coupled with multiple-cycle microwave treatments was identified as being more desirable for the carbonization processing." "Similarly, the PEI-functionalized hybrid dots PEI-C/Fe_3O_4-Dots were prepared by thermal carbonization processing of PEI-FAC mixtures in terms of the same multiple microwave treatment – quick cooling cycles, and isolated from the reaction mixture as the magnetically responsive fraction in a magnetic separation" (Hu et al. 2017).

X-ray powder diffraction patterns of the samples were due exclusively to Fe_3O_4 (Fig. 7.19), suggesting that the samples were free from any meaningful presence of other crystalline species, and the carbon domains in the hybrid dots were largely amorphous (Hu et al. 2017), as commonly observed for core carbon nanoparticles in CDots. "With the peak broadening, calculations based on the Scherrer equation yielded average Fe_3O_4 nanocrystal sizes of 10 ± 1 nm and 9.5 ± 0.9 nm in the PEG-C/Fe_3O_4-Dots and PEI-C/Fe_3O_4-Dots, respectively. The size estimates were supported by the TEM imaging results. Shown in Fig. 7.20 is a representative TEM image for the PEI-C/Fe_3O_4-Dots. Because of the high Z-contrast between carbon and iron, the size profiles in TEM images are dominated by the Fe_3O_4 nanoparticles in the hybrid dots. The statistical analysis based on multiple TEM images yielded an average size of 9.7 nm and a size distribution standard deviation of about 2 nm (Fig. 7.20), consistent with the estimate based on the Scherrer equation" (Hu et al. 2017).

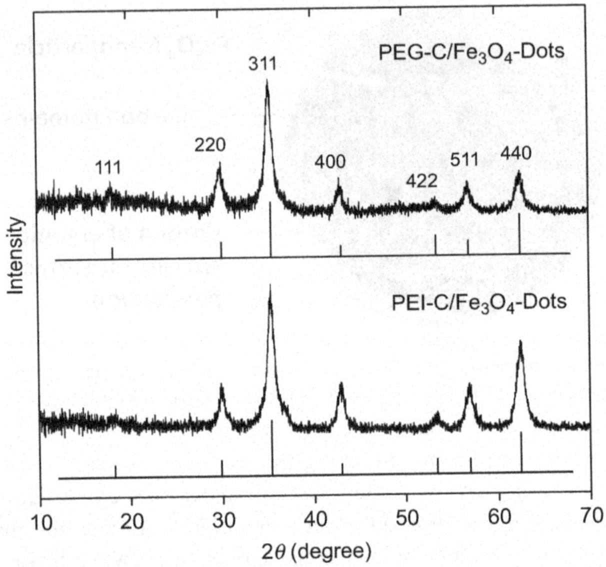

Fig. 7.19 Results from the X-ray powder diffraction characterization of the carbon/Fe_3O_4 hybrid dots, with the standard diffraction pattern of cubic Fe_3O_4 also shown for comparison. (Originally Figure 2 in Hu et al. (2017). Reproduced with permission)

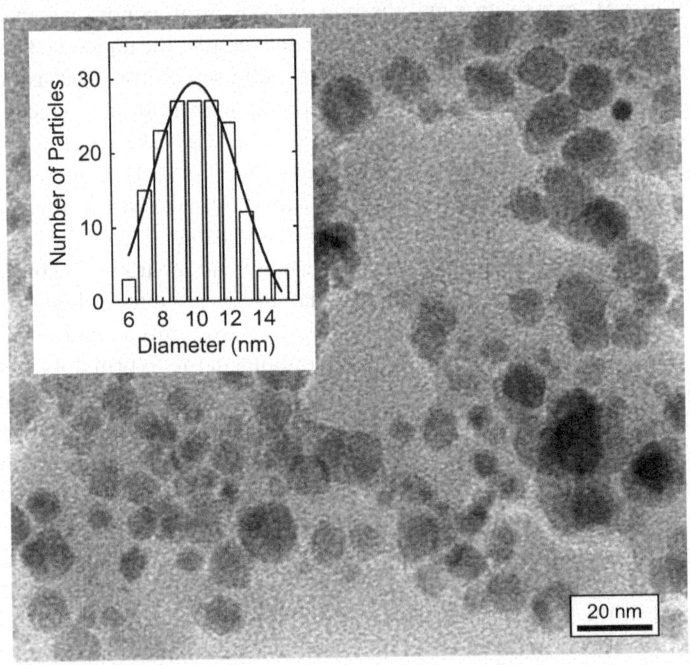

Fig. 7.20 A TEM image of the PEI-C/Fe_3O_4-Dots, with a statistical size analysis based on multiple images in the inset. (Originally Figure 3 in Hu et al. (2017). Reproduced with permission)

Aqueous dispersions of the hybrid dots exhibited more scattering in appearance visually and also in absorption spectral measurements even when the dispersions were very dilute, so that polyvinylpyrrolidone (PVP) polymers were added to reduce the light scattering in optical absorption measurements (Hu et al. 2017). "As shown in Fig. 7.21, the observed absorption spectra of the aqueous dispersed PEG-C/Fe_3O_4-Dots and PEI-C/Fe_3O_4-Dots with PVP polymers for improved dispersion are generally similar to that of the neat PEI-CDots. It should be pointed out that Fe_3O_4 nanoparticles are colored, but their absorptivities in the visible spectral region are generally low, and thus their contributions to the observed absorption spectra of the hybrid dots are insignificant" (Hu et al. 2017).

The average carbon content in the dot cores was estimated by using the known absorptivities of small carbon nanoparticles, and the iron content was determined by the classical colorimetric method based on the formation of red-colored iron(III) complex with thiocyanate (Hu et al. 2017). The estimated C/Fe mass ratios in the two hybrid dot samples were on the order of 0.5. Since the TEM results were against the possibility of carbon nanoparticles and Fe_3O_4 nanoparticles in pairs or clusters, a more likely dot configuration was a Fe_3O_4 nanoparticle surface-coated with a layer of carbon and/or decorated with randomly distributed small carbon domains, which were functionalized by residual PEG (or PEI) species or segments (Fig. 7.18). "The organic functionalization responsible for the passivation effect could be binding and/or the exposed segments of the organic species trapped in the dot surface structure. Such a hybrid dot configuration could be viewed in a first approximation as being on average a special structure of core and shells, namely the Fe_3O_4 nanoparticle core, a carbon shell (with an estimated thickness on the order of 1 nm), and then a soft-shell (or corona) of bound or attached organic species (Fig. 7.18). Obviously, this structural view may be too simplified and idealized, as the actual arrangements of the carbon and Fe_3O_4 domains in the hybrid dots are probably more random and vary from dot to dot" (Hu et al. 2017).

Fluorescence emissions of the aqueous dispersed PEG-C/Fe_3O_4-Dots and PEI-C/Fe_3O_4-Dots were similar in spectral profiles to those of neat CDots (Fig. 7.21), but the quantum yields (Φ_F) were lower. For example, the Φ_F of the more fluorescent PEI-C/Fe_3O_4-Dots on the order of 3% at 400 nm excitation was still much lower than those of the neat PEI-CDots (on average around 15% at the same excitation wavelength). "Since the optical absorptions and fluorescence emissions in the visible spectral region are associated with the carbon domains in the hybrid dots, the observed lower fluorescence yields could be attributed to causes including the different formation processes and structural configurations of the hybrid dots from those of their neat CDots counterparts (resulting in less effective passivation of the carbon domains) and/or direct quenching effect of the Fe_3O_4 in the hybrid dots on the emissive excited states of the carbon domains. The latter as a possible cause was examined experimentally by removing Fe_3O_4 from the PEI-C/Fe_3O_4-Dots, for which the dot sample was treated in an aqueous acid (HCl) solution to dissolve the iron oxide, followed by dialysis to harvest the residual nanostructures of the hybrid dots without the Fe_3O_4 nanoparticles. For these Fe_3O_4-free residual 'dots,' their observed fluorescence spectra are largely the same as those of the original PEI-C/Fe_3O_4-Dots

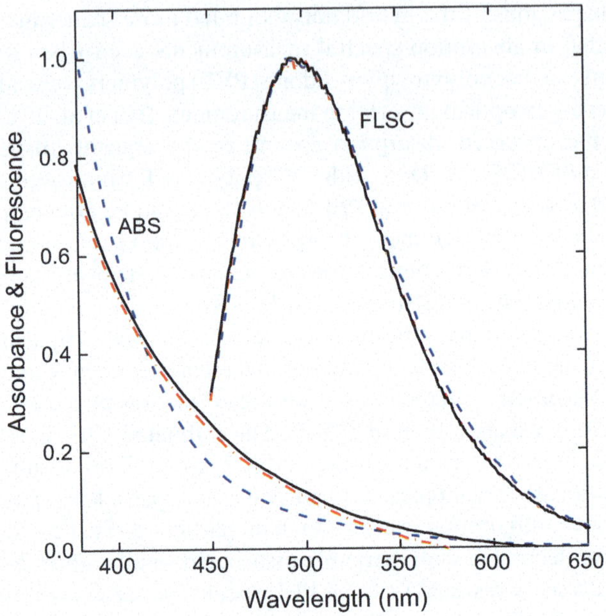

Fig. 7.21 Absorption (ABS) and fluorescence (FLSC, 440 nm excitation) spectra of the PEG-C/Fe_3O_4-Dots (-.-.-) and PEI-C/Fe_3O_4-Dots (——) in aqueous dispersions with PVP polymer as a dispersion agent are compared with those of the neat PEI-CDots in aqueous solution (- - - -). (Originally Figure 6 in Hu et al. (2017). Reproduced with permission)

before the acid treatment, again supporting the conclusion that the fluorescence emissions in the hybrid dots are due to the carbon domains passivated by the organic species. The fluorescence quantum yields of the hybrid dots post-acid treatment are around 4% at 400 nm excitation, somewhat higher than those of the same dots pre-acid treatment but still lower than those of the neat PEI-CDots reported previously. While there may be an argument that the comparison with the neat PEI-CDots is biased against the acid-treated hybrid dots because the acid treatment process may have other effects on the dot structures in addition to the removal of Fe_3O_4, such as effects that are negative to the effective passivation of the carbon domains by PEI and derived species in the residual dot structures post-acid treatment, it seems unlikely that the presence of Fe_3O_4 as a quencher within the hybrid dot structure could account entirely or predominantly for the observed lower fluorescence quantum yields of the hybrid dots. Therefore, it should be possible to prepare carbon/Fe_3O_4 hybrid dots of competitive fluorescence performances to those of their neat CDots counterparts by developing more suitable synthetic strategies" (Hu et al. 2017).

A major motivation in the development of carbon/Fe_3O_4 hybrid dots and related nanostructures has been for their serving as contrast agents in MRI and associated dual-/multi-modality imaging applications (e.g., Wang et al. 2014). For the same or similar purposes, carbon–gadolinium (Gd) hybrid dots or related nanostructures have been pursued (Bourlinos et al. 2012; Chen et al. 2014) with the general understanding that Gd-based MRI contrast agents are considered as superior in performance. Among the synthetic protocols have been the calcination and pyrolysis of organic mixtures with Gd in air and the hydrothermal treatment or brief microwave irradiation of similar mixtures. Relatively speaking, the calcination at a high temperature was one of only a few reported carbonization syntheses from which substantial contents of nanoscale carbon domains in the dot samples might be expected. In one study (Chen et al. 2014), the Gd-encapsulated "carbonaceous dots" (denoted as "Gd@C-dots") were prepared by using the popular Gd–DTPA complex as precursor for calcination at 300 °C in air for 2 h. The dots were evaluated for their photoluminescence properties and also serving as MRI contract agents (Fig. 7.22) (Chen et al. 2014).

The toxicity of Gd-based MRI contrast agents has attracted more recent attention and also serious concerns, as reflected by the warnings of the Food and Drug Administration (FDA). In the development of carbon–Gd hybrid dots for improved MRI performance, the potential Gd leakage has been a central issue, on which conflicting results have been presented in the literature. Certainly not helping the leakage problem have been the carbonization synthesis protocols in which the mild processing conditions would produce rather porous dot structures of crosslinked

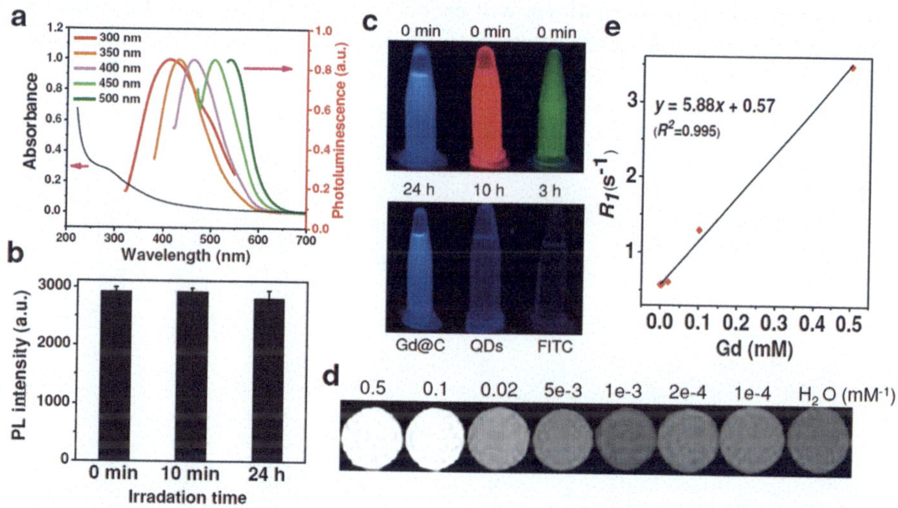

Fig. 7.22 Optical and magnetic properties of Gd@C-dots. (**a**) Absorption and photoluminescence spectra of Gd@C-dots. (**b**) Photostability study. Gd@C-dots were under continuous irradiation by UV light (30 W, 254 nm) and the photoluminescence (ex/em 360/425 nm) intensity was monitored over time. (**c**) Comparison of photostability among FITC, CdSe/ZnS QDs, and Gd@C-dots. The three solutions were under continuous irradiation by UV light (30 W, 254 nm) for different amounts of time. (**d**) T_1-weight MR images of Gd@C-dot agarose samples of different Gd concentrations. (**e**) Linear correlation between $R_1(T_1^{-1})$ and Gd concentration, based on readings from **d**. The r_1 relaxivity, which is the slope of the curve, was determined to be 5.88 mM^{-1} s^{-1}. (Originally Figure 2 in Chen et al. (2014). Reproduced with permission)

organic materials incorporated with some small carbon nano-domains. However, even with the much more robust carbonization processing conditions in the synthesis of the carbon/Gd hybrid dots, there are recent results (Wang et al. 2019) suggesting that the desired no Gd leakage may still represent a major challenge.

3 Concluding Remarks

Nanoscale carbon apparently has this amazing ability to "mix in" with other "things," which include molecules (encapsulated/embedded, crosslinked, and/or surface functionalized or decorated/conjugated), semiconductor nanoparticles, and nanoscale metals and insoluble metal salts in general. Carbon-based/carbon-derived hybrid nanostructures or hybrid CDots that exploit such an ability present essentially unlimited opportunities for both fundamental development (e.g., in terms of probing and understanding the interactions between the nanomaterials in the hybrid nanostructures) and an extremely broad range of applications. In fact, while CDots as a new class of QD-like nanomaterials have captured many of the unique or advantageous properties of small carbon nanoparticles, which represent the "carbon at zero-dimension" in the family of nanoscale carbon allotropes, one might envision that the carbon hybrid dot platform will enable substantially expanded exploration of the zero-dimensional carbon allotrope for more exciting discoveries and related technological advances.

References Cited

Abu Rabe, D.I., Al Awak, M.M., Yang, F., Okonjo, P.A., Dong, X., Teisl, L.R., Wang, P., Tang, Y., Pan, N., Sun, Y.-P., Yang, L.: The dominant role of surface functionalization in carbon dots' photo-activated antibacterial activity. Int. J. Nanomedicine. **14**, 2655–2665 (2019)

Al Awak, M.M., Wang, P., Wang, S., Tang, Y., Sun, Y.-P., Yang, L.: Correlation of carbon dots' light-activated antimicrobial activities and fluorescence quantum yields. RSC Adv. **7**, 177–30184 (2017)

Anilkumar, P., Wang, X., Cao, L., Sahu, S., Liu, J.-H., Wang, P., Korch, K., Tackett II, K.N., Parenzan, A., Sun, Y.-P.: Toward quantitatively fluorescent carbon-based "quantum" dots. Nanoscale. **3**, 2023–2027 (2011)

Anilkumar, P., Cao, L., Yu, J.-J., Tackett II, K.N., Wang, P., Meziani, M.J., Sun, Y.-P.: Crosslinked carbon dots as ultra-bright fluorescence probes. Small. **9**, 545–551 (2013)

Baker, M.: Nanotechnology imaging probes: smaller and more stable. Nat. Methods. **7**, 957–962 (2010)

Banerjee, S., Wong, S.S.: Structural characterization, optical properties, and improved solubility of carbon nanotubes functionalized with Wilkinson's catalyst. J. Am. Chem. Soc. **124**, 8940–8948 (2002)

Bourlinos, A.B., Bakandritsos, A., Kouloumpis, A., Gournis, D., Krysmann, M., Giannelis, E.P., Polakova, K., Safarova, K., Hola, K., Zboril, R.: Gd(III)-doped carbon dots as a dual florescent-MRI probe. J. Mater. Chem. **22**, 23327–23330 (2012)

Bunker, C.E., Lawson, G.E., Sun, Y.-P.: Fullerene-styrene random copolymers. Novel optical properties. Macromolecules. **28**, 3744–3746 (1995)

Cao, T., Webber, S.E.: Free radical copolymerization of styrene and C_{60}. Macromolecules. **29**, 3826–3830 (1996)

Cao, L., Wang, X., Meziani, M.J., Lu, F., Wang, H., Luo, P.G., Lin, Y., Harruff, B.A., Veca, L.M., Murray, D., Xie, S.-Y., Sun, Y.-P.: Carbon dots for multiphoton bioimaging. J. Am. Chem. Soc. **129**, 11318–11319 (2007)

Cao, L., Sahu, S., Anilkumar, P., Bunker, C.E., Xu, J., Fernando, K.A.S., Wang, P., Guliants, E.A., Tackett II, K.N., Sun, Y.-P.: Carbon nanoparticles as visible-light photocatalysts for efficient CO_2 conversion and beyond. J. Am. Chem. Soc. **133**, 4754–4757 (2011a)

Cao, L., Anilkumar, P., Wang, X., Liu, J.-H., Sahu, S., Meziani, M.J., Myers, E., Sun, Y.-P.: Reverse Stern-Volmer behavior for luminescence quenching in carbon nanoparticles. Can. J. Chem. **89**, 104–109 (2011b)

Cao, L., Yang, S.-T., Wang, X., Luo, P.G., Liu, J.-H., Sahu, S., Liu, Y., Sun, Y.-P.: Competitive performance of carbon "quantum" dots in optical bioimaging. Theranostics. **2**, 295–301 (2012)

Cao, L., Meziani, M.J., Sahu, S., Sun, Y.-P.: Photoluminescence properties of graphene versus other carbon nanomaterials. Acc. Chem. Res. **46**, 171–180 (2013)

© Springer Nature Switzerland AG 2020
Y.-P. Sun, *Carbon Dots*, https://doi.org/10.1007/978-3-030-41184-8

Cao, L., Fernando, K.A.S., Liang, W., Seilkop, A., Veca, L.M., Sun, Y.-P., Bunker, C.E.: Carbon dots for energy conversion applications. J. Appl. Phys. **125**(22), 220903 (2019)
Chattopadhyay, D., Galeska, I., Papadimitrakopoulos, F.: A route for bulk separation of semiconducting from metallic single-wall carbon nanotubes. J. Am. Chem. Soc. **125**, 3370–3375 (2003)
Chen, H., Wang, G.D., Tang, W., Todd, T., Zhen, Z., Tsang, C., Hekmatyar, K., Cowger, T., Hubbard, R.B., Zhang, W., Stickney, J., Shen, B., Xie, J.: Gd encapsulated carbonaceous dots with efficient renal clearance for magnetic resonance imaging. Adv. Mater. **26**, 6761–6766 (2014)
Cui, B., Yan, L., Gu, H., Yang, Y., Liu, X., Ma, C.-Q., Chen, Y., Jia, H.: Fluorescent carbon quantum dots synthesized by chemical vapor deposition: an alternative candidate for electron acceptor in polymer solar cells. Opt. Mater. **75**, 166–173 (2018)
Dong, X., Liang, W., Meziani, M.J., Sun, Y.-P., Yang, L.: Carbon dots as potent antimicrobial agents. Theranostics. **10**, 671–686 (2020)
Fernando, K.A.S., Sahu, S., Liu, Y., Lewis, W.K., Guliants, E.A., Jafariyan, A., Wang, P., Bunker, C.E., Sun, Y.-P.: Carbon quantum dots and applications in photocatalytic energy conversion. ACS Appl. Mater. Interfaces. **7**, 8363–8376 (2015)
Ford, W.T., Nishioka, T., McCleskey, S.C., Mourey, T.H., Kahol, P.: Structure and radical mechanism of formation of copolymers of C_{60} with styrene and with methyl methacrylate. Macromolecules. **33**, 2413–2423 (2000)
Fowler, P.W., Manolopoulos, D.E.: An Atlas of Fullerenes. Dover Publications, New York (2006). (ISBN 0486453626)
Gao, X., Du, C., Zhuang, Z., Chen, W.: Carbon quantum dot-based nanoprobes for metal ion detection. J. Mater. Chem. C. **4**, 6927–6945 (2016)
Ge, L., Pan, N., Jin, J., Wang, P., LeCroy, G.E., Liang, W., Yang, L., Teisl, L.R., Tang, Y., Sun, Y.-P.: Systematic comparison of carbon dots from different preparations – consistent optical properties and photoinduced redox characteristics in visible spectrum, and structural and mechanistic implications. J. Phys. Chem. C. **122**, 21667–21676 (2018)
Geerts, Y.H., Debever, O., Amato, C., Sergeyev, S.: Synthesis of mesogenic phthalocyanine-C_{60} donor–acceptor dyads designed for molecular heterojunction photovoltaic devices. Beilstein J. Org. Chem. **5**(49), 1–9 (2009). https://doi.org/10.3762/bjoc.5.49
Guldi, D.M., Holzinger, M., Hirsch, A., Georgakilasc, V., Prato, M.: First comparative emission assay of single-wall carbon nanotubes – solutions and dispersions. Chem. Commun. (10), 1130–1131 (2003)
Hewageegana, P., Apalkov, V.: Graphene quantum dots, Chapter 23. In: Sattler, K.D. (ed.) Handbook of Nanophysics: Functional Nanomaterials. CRC Press, Boca Raton (2017). (ISBN 9781138111936)
Hinterberger, V., Damm, C., Haines, P., Guldi, D.M., Peukert, W.: Purification and structural elucidation of carbon dots by column chromatography. Nanoscale. **11**, 8464–8474 (2019)
Hirsch, A., Brettreich, M.: Fullerenes: Chemistry and Reactions. Wiley-VCH Verlag GmbH & Co. KGaA, Weinheim (2005). (ISBN 3527308202)
Hou, X., Hu, Y., Wang, P., Yang, L., Al Awak, M.M., Tang, Y., Twara, F.K., Qian, H., Sun, Y.-P.: Modified facile synthesis for quantitatively fluorescent carbon dots. Carbon. **122**, 389–394 (2017)
Hu, Y., Al Awak, M.M., Yang, F., Yan, S., Xiong, Q., Wang, P., Tang, Y., Yang, L., LeCroy, G.E., Bunker, C.E., Xu, L., Tomlinson, N., Sun, Y.-P.: Photoexcited state properties of carbon dots from thermally induced functionalization of carbon nanoparticles. J. Mater. Chem. C. **4**, 10554–10561 (2016)
Hu, Y., Wang, P., Bunker, C.E., Teisl, L.R., Reibold, M., Yan, S., Qian, H., He, D., Sun, Y.-P.: Preparation and optical properties of magnetic carbon/iron oxide hybrid dots. RSC Adv. **7**, 41304–41310 (2017)
Huang, W., Taylor, S., Fu, K., Lin, Y., Zhang, D., Hanks, T., Rao, A.M., Sun, Y.-P.: Attaching proteins to carbon nanotubes via diimide-activated amidation. Nano Lett. **2**, 311–314 (2002)

Huang, X., Zhang, F., Zhu, L., Choi, K.Y., Guo, N., Guo, J., Tackett, K., Anilkumar, P., Liu, G., Quan, Q., Choi, H.S., Niu, G., Sun, Y.-P., Lee, S., Chen, X.: Effect of injection routes on the biodistribution, clearance, and tumor uptake of carbon dots. ACS Nano. **7**, 5684–5693 (2013)

Jang, J., Oh, J.H., Stucky, G.D.: Fabrication of ultrafine conducting polymer and graphite nanoparticles. Angew. Chem. Int. Ed. **41**, 4016–4019 (2002)

Juzenas, P., Chen, W., Sun, Y.-P., Coelho, M.A.N., Generalov, R., Generalova, N., Christensen, I.L.: Quantum dots and nanoparticles for photodynamic and radiotherapy of cancer. Adv. Drug Deliv. Rev. **60**, 1600–1614 (2008)

Juzenas, P., Kleinauskas, A., Luo, P.G., Sun, Y.-P.: Photoactivatable carbon nanodots for cancer therapy. Appl. Phys. Lett. **103**, 063701 (2013)

Khan, S., Sharma, A., Ghoshal, S., Jain, S., Hazra, M.K., Nandi, C.K.: Small molecular organic nanocrystals resemble carbon nanodots in terms of their properties. Chem. Sci. **9**, 175–180 (2018)

Kingston, C., Simard, B.: Fabrication of carbon nanotubes. Anal. Lett. **36**, 3119–3145 (2003)

Kleinauskas, A., Rocha, S., Sahu, S., Sun, Y.-P., Juzenas, P.: Carbon-core silver-shell nanodots as sensitizers for phototherapy and radiotherapy. Nanotechnology. **24**, 325103 (2013)

Kuo, W.S., Chang, C.Y., Chen, H.H., Hsu, C.L.L., Wang, J.Y., Kao, H.F., Chou, L.C.-S., Chen, Y.-C., Chen, S.-J., Chang, W.-T., Tseng, S.W., Wu, P.-C., Pu, Y.-C.: Two-photon photoexcited photodynamic therapy and contrast agent with antimicrobial graphene quantum dots. ACS Appl. Mater. Interfaces. **8**, 30467–30474 (2016)

LeCroy, G.E., Sonkar, S.K., Yang, F., Veca, L.M., Wang, P., Tackett II, K.N., Yu, J.-J., Vasile, E., Qian, H., Liu, Y., Luo, P.G., Sun, Y.-P.: Toward structurally defined carbon dots as ultracompact fluorescent probes. ACS Nano. **8**, 4522–4529 (2014)

LeCroy, G.E., Fernando, K.A.S., Bunker, C.E., Wang, P., Tomlinson, N., Sun, Y.-P.: Steady-state and time-resolved fluorescence studies on interactions of carbon "quantum" dots with nitrotoluenes. Inorg. Chim. Acta. **468**, 300–307 (2017a)

LeCroy, G.E., Messina, F., Sciortino, A., Bunker, C.E., Wang, P., Fernando, K.A.S., Sun, Y.-P.: Characteristic excitation wavelength dependence of fluorescence emissions in carbon "quantum" dots. J. Phys. Chem. C. **121**, 28180–28186 (2017b)

LeCroy, G.E., Wang, P., Bunker, C.E., Fernando, K.A.S., Liang, W., Ge, L., Reibold, M., Sun, Y.-P.: Hybrid carbon dots platform enabling opportunities for desired optical properties and redox characteristics by-design. Chem. Phys. Lett. **724**, 8–12 (2019)

Leménager, G., De Luca, E., Sun, Y.-P., Pompa, P.P.: Super-resolution fluorescence imaging of biocompatible carbon dots. Nanoscale. **6**, 8617–8623 (2014)

Li, H., Zhou, B., Lin, Y., Gu, L., Allard, L.F., Sun, Y.-P.: Selective interactions of porphyrins with semiconducting single-walled carbon nanotubes. J. Am. Chem. Soc. **126**, 1014–1015 (2004)

Li, H., Huang, W., Guo, Z., Guduru, R., Sun, Y.-P.: Derivatized fullerenes bearing multiple electron donors: synthesis and charge transfer properties. J. Phys. Chem. B. **109**, 11886–11892 (2005a)

Li, H., Kitaygorodskiy, A., Carino, R., Sun, Y.-P.: Simple modification in hexakis-addition for efficient synthesis of C60-centered dendritic molecules bearing multiple aromatic chromophores. Org. Lett. **7**, 859–861 (2005b)

Li, H., Haque, S.A., Kitaygorodskiy, A., Meziani, M.J., Torres-Castillo, M., Sun, Y.-P.: Alternatively modified bingel reaction for efficient syntheses of C_{60} hexakis-adducts. Org. Lett. **8**, 5641–5643 (2006)

Liang, W., Sun, Y.-P.: Red/near-IR organic dyes, not "carbon quantum dots". Unpublished results (2019)

Liang, W., Ge, L., Hou, X., Ren, X., Yang, L., Bunker, C.E., Overton, C.M., Wang, P., Sun, Y.-P.: Evaluation of commercial "carbon quantum dots" sample on origins of red absorption and emission features. C – J. Carbon Res. **5**, 70 (2019)

Lin, Y., Zhou, B., Fernando, K.A.S., Liu, P., Allard, L.F., Sun, Y.-P.: Polymeric carbon nanocomposites from carbon nanotubes functionalized with matrix polymer. Macromolecules. **36**, 7199–7204 (2003)

Lin, Y., Zhou, B., Martin, R.B., Henbest, K.B., Harruff, B.A., Riggs, J.E., Guo, Z.-X., Allard, L.F., Sun, Y.-P.: Visible luminescence of carbon nanotubes and dependence on functionalization. J. Phys. Chem. B. **109**, 14779–14782 (2005)

Liu, J.-H., Yang, S.-T., Wang, X., Wang, H., Liu, Y., Luo, P.G., Liu, Y., Sun, Y.-P.: Carbon nanoparticles trapped in vivo – similar to carbon nanotubes in time-dependent biodistribution. ACS Appl. Mater. Interfaces. **6**, 14672–14678 (2014)

Liu, J.-H., Cao, L., LeCroy, G.E., Wang, P., Meziani, M.J., Dong, Y., Liu, Y., Sun, Y.-P.: Fluorescence labeling of cells with carbon "quantum" dots. ACS Appl. Mater. Interfaces. **7**, 19439–19445 (2015)

Liu, Y., Wang, P., Fernando, K.A.S., LeCroy, G.E., Maimaiti, H., Harruff-Miller, B.A., Lewis, W.K., Bunker, C.E., Hou, Z.-L., Sun, Y.-P.: Enhanced fluorescence properties of carbon dots in polymer films. J. Mater. Chem. C. **4**, 6967–6974 (2016)

Liu, Y., Liu, Y., Qian, H., Wang, P., LeCroy, G.E., Bunker, C.E., Fernando, K.A.S., Yang, L., Reibold, M., Sun, Y.-P.: Carbon-TiO_2 hybrid dots in different configurations – optical properties, redox characteristics, and mechanistic implications. New J. Chem. **42**, 10798–10806 (2018)

Liu, J.-H., Wang, Y., Yang, G.-H., Yang, F., Gao, H., Huang, Y., Wang, H., Wang, P., Yang, L., Tang, Y., Teisl, L.R., Sun, Y.-P.: Systematic toxicity evaluations of high-performance carbon "quantum" dots. J. Nanosci. Nanotechnol. **19**, 2130–2137 (2019)

Lu, F., Meziani, M.J., Cao, L., Sun, Y.-P.: Separated metallic and semiconducting single-walled carbon nanotubes – opportunities in transparent electrodes and beyond. Langmuir. **27**, 4339–4350 (2011)

Ma, B., Sun, Y.-P.: Fluorescence spectra and quantum yields of [60]fullerene and [70]fullerene under different solvent conditions. A quantitative examination using a near-infrared-sensitive emission spectrometer. J. Chem. Soc. Perkin Trans. **2**, 2157–2162 (1996)

Ma, B., Bunker, C.E., Guduru, R., Zhang, X.-F., Sun, Y.-P.: Quantitative spectroscopic studies of the photoexcited state properties of methano- and pyrrolidino-[60]fullerene derivatives. J. Phys. Chem. A. **101**, 5626–5632 (1997)

Martin, R.B., Fu, K., Li, H., Cole, D., Sun, Y.-P.: Interesting fluorescence properties of C_{60}-centered dendritic adduct with twelve symmetrically attached pyrenes. Chem. Commun. (18), 2368–2369 (2003)

Meziani, M.J., Dong, X., Zhu, L., Jones, L.P., LeCroy, G.E., Yang, F., Wang, P., Zhao, Y., Yang, L., Tripp, R.A., Sun, Y.-P.: Visible-light-activated bactericidal functions of carbon "quantum" dots. ACS Appl. Mater. Interfaces. **8**, 10761–10766 (2016)

Nagasawa, N., Sugiyama, H., Naka, N., Kudryashov, I., Watanabe, M., Hayashi, T., Bozovic, I., Bozovic, N., Li, G., Li, Z., Tang, Z.K.: Visible emission of single-wall carbon nanotubes formed in micro-channels of zeolite crystals. J. Lumin. **97**, 161–167 (2002)

Niyogi, S., Hu, H., Hamon, M.A., Bhowmik, P., Zhao, B., Rozenzhak, S.M., Chen, J., Itkis, M.E., Meier, M.S., Haddon, R.C.: Chromatographic purification of soluble single-walled carbon nanotubes (s-SWNTs). J. Am. Chem. Soc. **123**, 733–734 (2001)

Pan, N., Okonjo, P.A., Wang, P., Tang, Y., Sun, Y.-P., Yang, L.: Optical and photodynamic properties of carbon/TiO2 hybrid dots in different nanoscale configurations. Chem. Phys. Lett. **743**, article 137208 (2020). https://doi.org/10.1016/j.cplett.2020.137208

Ponomarenko, L.A., Schedin, F., Katsnelson, M.I., Yang, R., Hill, E.W., Novoselov, K.S., Geim, A.K.: Chaotic Dirac billiard in graphene quantum dots. Science. **320**, 356–358 (2008)

Rance, G.A., Marsh, D.H., Nicholas, R.J., Khlobystov, A.N.: UV–vis absorption spectroscopy of carbon nanotubes: relationship between the π-electron plasmon and nanotube diameter. Chem. Phys. Lett. **493**, 19–23 (2010)

Rednic, M.I., Lu, Z., Wang, P., LeCroy, G.E., Yang, F., Liu, Y., Qian, H., Terec, A., Veca, L.M., Lu, F., Sun, Y.-P.: Versatile synthesis of fluorescent carbon dots from thermochemical functionalization of carbon nanoparticles. Chem. Phys. Lett. **639**, 109–113 (2015)

Ren, X., Liang, W., Wang, P., Bunker, C.E., Coleman, M., Teisl, L.R., Cao, L., Sun, Y.-P.: A new approach in functionalization of carbon nanoparticles for optoelectronically relevant carbon dots and beyond. Carbon. **141**, 553–560 (2019)

Riggs, J.E., Guo, Z., Carroll, D.L., Sun, Y.-P.: Strong luminescence of solubilized carbon nanotubes. J. Am. Chem. Soc. **122**, 5879–5880 (2000)

Robertson, J.: Amorphous carbon. Adv. Phys. **35**, 317–374 (1986)

Sahu, S., Liu, Y., Wang, P., Bunker, C.E., Fernando, K.A.S., Lewis, W.K., Guliants, E.A., Wang, J., Sun, Y.-P.: Visible-light photoconversion of carbon dioxide into organic acids in an aqueous solution of carbon dots. Langmuir. **30**, 8631–8636 (2014)

Salazar, M.A., Li, H., Sun, Y.-P., McNeill, J.: Spatial distribution of functional groups on shortened single-walled carbon nanotubes determined by near-field scanning optical microscopy. Unpublished results (2005)

Sciortino, A., Madonia, A., Gazzetto, M., Sciortino, L., Rohwer, E.J., Feurer, T., Gelardi, F.M., Cannas, M., Cannizzo, A., Messina, F.: The interaction of photoexcited carbon nanodots with metal ions disclosed down to the femtosecond scale. Nanoscale. **9**, 11902–11911 (2017)

Suda, Y., Ono, T., Akazawa, M., Sakai, Y., Tsujino, J., Homma, N.: Preparation of carbon nanoparticles by plasma-assisted pulsed laser deposition method – size and binding energy dependence on ambient gas pressure and plasma condition. Thin Solid Films. **415**, 15–20 (2002)

Sun, Y.-P.: Fluorescent carbon nanoparticles. U.S. Patent 7,829,772 B2 (2010)

Sun, Y.-P., Bunker, C.E.: C70 in solvent mixtures. Nature. **365**, 398 (1993)

Sun, Y.-P., Bunker, C.E., Ma, B.: Quantitative studies of ground and excited state charge transfer complexes of fullerenes with *N,N*-dimethylaniline and *N,N*-diethylaniline. J. Am. Chem. Soc. **116**, 9692 (1994)

Sun, Y.-P., Ma, B., Bunker, C.E., Liu, B.: All-carbon polymers (polyfullerenes) from photochemical reactions of fullerene clusters in room-temperature solvent mixtures. J. Am. Chem. Soc. **117**, 12705–12711 (1995)

Sun, Y.-P., Lawson, G.E., Riggs, J.E., Ma, B., Wang, N., Moton, D.K.: Photophysical and nonlinear optical properties of [60]fullerene derivatives. J. Phys. Chem. A. **102**, 5520–5528 (1998)

Sun, Y.-P., Guduru, R., Lawson, G.E., Mullins, J.E., Guo, Z., Bunker, C.E., Gord, J.R., Quinlan, J.: Photophysical and electron transfer properties of mono- and multiple-functionalized fullerene derivatives. J. Phys. Chem. B. **104**, 4625–4632 (2000)

Sun, Y.-P., Zhou, B., Henbest, K., Huang, W., Fu, K., Lin, Y., Taylor, S., Carroll, D.L.: Luminescence anisotropy of functionalized carbon nanotubes in solution. Chem. Phys. Lett. **351**, 349–353 (2002a)

Sun, Y.-P., Huang, W., Guduru, R., Martin, R.B.: Intramolecular electron transfer in fullerene derivatives with multiple donors. Chem. Phys. Lett. **353**, 353–358 (2002b)

Sun, Y.-P., Zhou, B., Lin, Y., Wang, W., Fernando, K.A.S., Pathak, P., Meziani, M.J., Harruff, B.A., Wang, X., Wang, H., Luo, P.G., Yang, H., Kose, M.E., Chen, B., Veca, L.M., Xie, S.-Y.: Quantum-sized carbon dots for bright and colorful photoluminescence. J. Am. Chem. Soc. **128**, 7756–7757 (2006)

Sun, Y.-P., Wang, X., Lu, F., Cao, L., Meziani, M.J., Luo, P.G., Gu, L., Veca, L.M.: Doped carbon nanoparticles as new platform for highly photoluminescent dots. J. Phys. Chem. C. **112**, 18295–18298 (2008)

Sun, Y.-P., Wang, P., Lu, Z., Yang, F., Meziani, M.J., LeCroy, G.E., Liu, Y., Qian, H.: Host-guest carbon dots for enhanced optical properties and beyond. Sci. Rep. **5**, 12354 (2015). https://doi.org/10.1038/srep12354

Veca, L.M., Lu, F., Meziani, M.J., Cao, L., Zhang, P., Shrestha, M., Sun, Y.-P.: Polymer functionalization and solubilization of carbon nanosheets. Chem. Commun. (18), 2565–2567 (2009)

Veca, L.M., Diac, A., Mihalache, I., Wang, P., LeCroy, G.E., Pavelescu, E., Gavrila, R., Vasile, E., Terec, A., Sun, Y.-P.: Electroluminescence of carbon "quantum" dots – from materials to devices. Chem. Phys. Lett. **613**, 40–44 (2014)

Wang, Y.: Nonlinear optical properties of fullerenes and charge-transfer complexes of fullerenes. J. Phys. Chem. **96**, 1530–1532 (1992)

Wang, X.: Preparation and studies of fluorescent carbon nanomaterials. PhD Dissertation, Clemson University (2010)

Wang, X., Cao, L., Lu, F., Meziani, M.J., Li, H., Qi, G., Zhou, B., Harruff, B.A., Sun, Y.-P.: Photoinduced electron transfers with carbon dots. Chem. Commun. (25), 3774–3776 (2009)

Wang, X., Cao, L., Yang, S.-T., Lu, F., Meziani, M.J., Tian, L., Sun, K.W., Bloodgood, M.A., Sun, Y.-P.: Bandgap-like strong fluorescence in functionalized carbon nanoparticles. Angew. Chem. Int. Ed. **49**, 5310–5314 (2010a)

Wang, X., Cao, L., Bunker, C.E., Meziani, M.J., Lu, F., Guliants, E.A., Sun, Y.-P.: Fluorescence decoration of defects in carbon nanotubes. J. Phys. Chem. C. **114**, 20941–20946 (2010b)

Wang, Y., Anilkumar, P., Cao, L., Liu, J.-H., Luo, P.G., Tackett II, K.N., Sahu, S., Wang, P., Wang, X., Sun, Y.-P.: Carbon dots of different composition and surface functionalization – cytotoxicity issues relevant to fluorescence imaging of cells. Exp. Biol. Med. **236**, 1231–1238 (2011)

Wang, J., Sahu, S., Sonkar, S.K., Tackett II, K.N., Sun, K.W., Liu, Y., Anilkumar, P., Sun, Y.-P.: Versatility with carbon dots – from overcooked BBQ to brightly fluorescent agents and photocatalysts. RSC Adv. **3**, 15604–15607 (2013)

Wang, H., Shen, J., Li, Y., Wei, Z., Cao, G., Gai, Z., Hong, K., Banerjee, P., Zhou, S.: Magnetic iron oxide – fluorescent carbon dots integrated nanoparticles for dual-modal imaging, near-infrared light-responsive drug carrier and photothermal therapy. Biomater. Sci. **2**, 915–923 (2014)

Wang, P., Liu, J.-H., Gao, H., Hu, Y., Hou, X., LeCroy, G.E., Liang, W., Chen, J., Bunker, C.E., Liu, Y., Sun, Y.-P.: Host-guest carbon dots as high-performance fluorescence probes. J. Mater. Chem. C. **5**, 6328–6335 (2017)

Wang, P., Ren, X., Sun, Y.-P.: Preparation and properties of carbon-gadolinium hybrid dots. Unpublished results (2019)

Westervelt, R.M.: Graphene nanoelectronics. Science. **320**, 324–325 (2008)

Xie, Q., Perez-Cordero, E., Echegoyen, L.: Electrochemical detection of C_{60}^{6-} and C_{70}^{6-}: enhanced stability of fullerides in solution. J. Am. Chem. Soc. **114**, 3978–3980 (1992)

Xiong, Y., Schneider, J., Ushakova, E.V., Rogach, A.L.: Influence of molecular fluorophores on the research field of chemically synthesized carbon dots. Nano Today. **23**, 124–139 (2018)

Xu, J., Sahu, S., Cao, L., Anilkumar, P., Tackett II, K.N., Qian, H., Bunker, C.E., Guliants, E.A., Parenzan, A., Sun, Y.-P.: Carbon nanoparticles as "chromophores" for photon harvesting and photoconversion. ChemPhysChem. **12**, 3604–3608 (2011)

Xu, J., Sahu, S., Cao, L., Bunker, C.E., Peng, G., Fernando, K.A.S., Wang, P., Guliants, E.A., Meziani, M.J., Qian, H., Sun, Y.-P.: Efficient fluorescence quenching in carbon dots by surface-doped metals – disruption of excited state redox processes and mechanistic implications. Langmuir. **28**, 16141–16147 (2012)

Yan, L., Yang, Y., Ma, C.-Q., Liu, X., Wang, H., Xu, B.: Synthesis of carbon quantum dots by chemical vapor deposition approach for use in polymer solar cell as the electrode buffer layer. Carbon. **109**, 598–607 (2016)

Yang, S.-T., Ho, Y.-S.: Research performance and trends of fluorescent carbon nanoparticles: a science citation index expanded-based analysis. J. Nanopart. Res. **21**, 202 (2019)

Yang, S.-T., Guo, W., Lin, Y., Deng, X.-Y., Wang, H.-F., Sun, H.-F., Liu, Y.-F., Wang, X., Wang, W., Chen, M., Huang, Y.-P., Sun, Y.-P.: Biodistribution of pristine single-walled carbon nanotubes in vivo. J. Phys. Chem. C. **111**, 17761–17764 (2007)

Yang, S.-T., Fernando, K.A.S., Liu, J.-H., Wang, J., Sun, H.-F., Liu, Y.-F., Chen, M., Huang, Y.-P., Wang, X., Wang, H.-F., Sun, Y.-P.: Covalently PEGylated carbon nanotubes with stealth character in vivo. Small. **4**, 940–944 (2008)

Yang, S.-T., Cao, L., Luo, P.G., Lu, F., Wang, X., Wang, H., Meziani, M.J., Liu, Y., Qi, G., Sun, Y.-P.: Carbon dots for optical imaging in vivo. J. Am. Chem. Soc. **131**, 11308–11309 (2009a)

Yang, S.-T., Wang, H., Meziani, M.J., Liu, Y., Wang, X., Sun, Y.-P.: Bio-defunctionalization of functionalized single-walled carbon nanotubes in mice. Biomacromolecules. **10**, 2009–2012 (2009b)

Yang, S.-T., Wang, X., Wang, H., Lu, F., Luo, P.G., Cao, L., Meziani, M.J., Liu, J.-H., Liu, Y., Chen, M., Huang, Y., Sun, Y.-P.: Carbon dots as nontoxic and high-performance fluorescence imaging agents. J. Phys. Chem. C. **113**, 18110–18114 (2009c)

Yang, F., LeCroy, G.E., Wang, P., Liang, W., Chen, J., Fernando, K.A.S., Bunker, C.E., Qian, H., Sun, Y.-P.: Functionalization of carbon nanoparticles and defunctionalization – towards structural and mechanistic elucidation of carbon "quantum" dots. J. Phys. Chem. C. **120**, 25604–25611 (2016)

Yang, S.-T., Lin, J.-H., Wang, P., Yang, S., Ge, L., Yan, S., Sun, Y.-P.: High-performance red/near-IR carbon dots as fluorescence probes for tumor imaging in vivo. ChemistrySelect. **3**, 6374–6381 (2018a)

Yang, F., Ren, X., LeCroy, G.E., Song, J., Wang, P., Beckerle, L., Bunker, C.E., Xiong, Q., Sun, Y.-P.: Zero-dimensional carbon allotropes – carbon nanoparticles versus fullerenes in functionalization by electronic polymers for different optical and redox properties. ACS Omega. **3**, 5685–5691 (2018b)

Zhang, X., Zhang, Y., Wang, Y., Kalytchuk, S., Kershaw, S.V., Wang, Y., Wang, P., Zhang, T., Zhao, Y., Zhang, H., Cui, T., Wang, Y., Zhao, J., Yu, W.W., Rogach, A.L.: Color-switchable electroluminescence of carbon dot light-emitting diodes. ACS Nano. **7**, 11234–11241 (2013)

Zhao, B., Hu, H., Niyogi, S., Itkis, M.E., Hamon, M.A., Bhowmik, P., Meier, M.S., Haddon, R.C.: Chromatographic purification and properties of soluble single-walled carbon nanotubes. J. Am. Chem. Soc. **123**, 11673–11677 (2001)

Zhou, B., Lin, Y., Veca, L.M., Fernando, K.A.S., Sun, Y.-P.: Study of electronic transitions in carbon nanotubes by luminescence polarization spectroscopy in polymeric matrix. J. Phys. Chem. B. **110**, 3001–3006 (2006)

Zhou, J., Booker, C., Li, R., Zhou, X., Sham, T.-K., Sun, X., Ding, Z.: An electrochemical avenue to blue luminescent nanocrystals from multiwalled carbon nanotubes (MWCNTs). J. Am. Chem. Soc. **129**, 744–745 (2007)

Bibliography – A List of Reviews

Atabaev, T.S.: Doped carbon dots for sensing and bioimaging applications: a minireview. Nanomater. **8**, 342 (2018)

Cao, L., Meziani, M.J., Sahu, S., Sun, Y.-P.: Photoluminescence properties of graphene versus other carbon nanomaterials. Acc. Chem. Res. **46**, 171–180 (2013)

Cao, L., Fernando, K.A.S., Liang, W., Seilkop, A., Veca, L.M., Sun, Y.-P., Bunker, C.E.: Carbon dots for energy conversion applications. J. Appl. Phys. **125**(22), 220903 (2019)

Chen, W., Lv, G., Hu, W., Li, D., Chen, S., Dai, Z.: Synthesis and application of graphene quantum dots: a review. Nanotechnol. Rev. **7**, 157–185 (2018)

Das, R., Bandyopadhyay, R., Pramanik, P.: Carbon quantum dots from natural resource: a review. Mater. Today Chem. **8**, 96–109 (2018)

Devi, P., Rajput, P., Thakur, A., Kim, K.-H., Kumar, P.: Recent advances in carbon quantum dot-based sensing of heavy metals in water. Trends Anal. Chem. **114**, 171–195 (2019)

Dong, X., Liang, W., Meziani, M.J., Sun, Y.-P., Yang, L.: Carbon dots as potent antimicrobial agents. Theranostics. **10**, 671–686 (2020)

Essner, J.B., Baker, G.A.: The emerging roles of carbon dots in solar photovoltaics: a critical review. Environ. Sci. Nano. **4**, 1216–1263 (2017)

Fernando, K.A.S., Sahu, S., Liu, Y., Lewis, W.K., Guliants, E.A., Jafariyan, A., Wang, P., Bunker, C.E., Sun, Y.-P.: Carbon quantum dots and applications in photocatalytic energy conversion. ACS Appl. Mater. Interfaces. **7**, 8363–8376 (2015)

Gao, X., Du, C., Zhuang, Z., Chen, W.: Carbon quantum dot-based nanoprobes for metal ion detection. J. Mater. Chem. C. **4**, 6927–6945 (2016)

Gao, D., Zhao, H., Chen, X., Fan, H.: Recent advance in red-emissive carbon dots and their photoluminescent mechanisms. Mater. Today. **9**, 103–113 (2018)

Georgakilas, V., Perman, J.A., Tucek, J., Zboril, R.: Broad family of carbon nanoallotropes: classification, chemistry, and applications of fullerenes, carbon dots, nanotubes, graphene, nanodiamonds, and combined superstructures. Chem. Rev. **115**, 4744–4822 (2015)

Hassan, M., Gomes, V.G., Dehghani, A., Ardekani, S.M.: Engineering carbon quantum dots for photomediated theranostics. Nano Res. **11**, 1–41 (2018)

Hoang, V.C., Dave, K., Gomes, V.G.: Carbon quantum dot-based composites for energy storage and electrocatalysis: mechanisms, applications, and future prospects. Nano Energy. **66**, 104093 (2019)

Hu, C., Li, M., Qiu, J., Sun, Y.-P.: Design and fabrication of carbon dots for energy conversion and storage. Chem. Soc. Rev. **48**, 2315–2337 (2019)

Hutton, G.A.M., Martindale, B.C.M., Reisner, E.: Carbon dots as photosensitisers for solar-driven catalysis. Chem. Soc. Rev. **46**, 6111–6123 (2017)

© Springer Nature Switzerland AG 2020

Y.-P. Sun, *Carbon Dots*, https://doi.org/10.1007/978-3-030-41184-8

Kaur, M., Kaur, M., Sharma, V.K.: Nitrogen-doped graphene and graphene quantum dots: a review on synthesis and application in energy, sensors, and environment. Adv. Colloid Interf. Sci. **259**, 44–64 (2018)

Kovacova, M., Spitalska, E., Markovic, Z., Spitalsky, Z.: Carbon quantum dots as antibacterial photosensitizers and their polymeric nanocomposite applications. Part. Part. Syst. Charact. **37**, 1900348 (2020)

LeCroy, G.E., Yang, S.-T., Yang, F., Liu, Y., Fernando, K.A.S., Bunker, C.E., Hu, Y., Luo, P.G., Sun, Y.-P.: Functionalized carbon nanoparticles: syntheses and applications in optical bioimaging and energy conversion. Coord. Chem. Rev. **320**, 66–81 (2016)

Li, X., Rui, M., Song, J., Shen, Z., Zeng, H.: Carbon and graphene quantum dots for optoelectronic and energy devices: a review. Adv. Funct. Mater. **25**, 4929–4947 (2015)

Li, M., Chen, T., Gooding, J.J., Liu, J.: Review of carbon and graphene quantum dots for sensing. ACS Sens. **4**, 1732–1748 (2019)

Liang, W., Bunker, C.E., Sun, Y.-P.: Carbon dots: zero-dimensional carbon allotrope with unique photoinduced redox characteristics. ACS Omega. **5**(2), 965–971 (2020)

Lim, S.Y., Shen, W., Gao, Z.: Carbon quantum dots and their applications. Chem. Soc. Rev. **44**, 362–381 (2015)

Liu, M.: Optical properties of carbon dots: a review. Nanoarchitectonics. **1**, 1–12 (2020)

Liu, M.L., Chen, B.B., Li, C.M., Huang, C.Z.: Carbon dots: synthesis, formation mechanism, fluorescence origin and sensing applications. Green Chem. **21**, 449–471 (2019)

Luo, P.G., Sahu, S., Yang, S.-T., Sonkar, S.K., Wang, J., Wang, H., LeCroy, G.E., Cao, L., Sun, Y.-P.: Carbon "quantum" dots for optical bioimaging. J. Mater. Chem. B. **1**, 2116–2127 (2013)

Luo, P.G., Sonkar, S.K., Yang, S.-T., Yang, F., Yang, L., Broglie, J.J., Sun, Y.-P.: Carbon-based quantum dots for fluorescence imaging of cells and tissues. RSC Adv. **4**, 10791–10807 (2014)

Maiti, D., Tong, X., Mou, X., Yang, K.: Carbon-based nanomaterials for biomedical applications: a recent study. Front. Pharmacol. **9**, 1401 (2019)

Mehta, A., Mishra, A., Basu, S., Shetti, N.P., Reddy, K.R., Saleh, T.A., Aminabhavi, T.M.: Band gap tuning and surface modification of carbon dots for sustainable environmental remediation and photocatalytic hydrogen production – a review. J. Environ. Manag. **250**, 109486 (2019)

Molaei, M.J.: A review on nanostructured carbon quantum dots and their applications in biotechnology, sensors, and chemiluminescence. Talanta. **196**, 456–478 (2019a)

Molaei, M.J.: Carbon quantum dots and their biomedical and therapeutic applications: a review. RSC Adv. **9**, 6460–6481 (2019b)

Namdari, P., Negahdari, B., Eatemadi, A.: Synthesis, properties and biomedical applications of carbon-based quantum dots: an updated review. Biomed. Pharmacother. **87**, 209–222 (2017)

Pan, M., Yin, Z., Liu, K., Du, X., Liu, H., Wang, S.: Carbon-based nanomaterials in sensors for food safety. Nanomater. **9**, 1330 (2019)

Pardo, J., Peng, Z., Leblanc, R.M.: Cancer targetting and drug delivery using carbon-based quantum dots and nanotubes. Molecules. **23**, 378 (2018)

Peng, Z., Han, X., Li, S., Al-Youbi, A.O., Bashammakh, A.S., El-Shahawi, M.S., Leblanc, R.M.: Carbon dots: biomacromolecule interaction, bioimaging and nanomedicine. Coord. Chem. Rev. **343**, 256–277 (2017)

Pirsaheb, M., Mohammadi, S., Salimi, A., Payandeh, M.: Functionalized fluorescent carbon nanostructures for targeted imaging of cancer cells: a review. Microchim. Acta. **186**, 231 (2019)

Roy, P., Chen, P.-C., Periasamy, A.P., Chen, Y.-N., Chang, H.-T.: Photoluminescent carbon nanodots: synthesis, physicochemical properties and analytical applications. Mater. Today. **18**, 447–458 (2015)

Semeniuk, M., Yi, Z., Poursorkhabi, V., Tjong, J., Jaffer, S., Lu, Z.-H., Sain, M.: Future perspectives and review on organic carbon dots in electronic applications. ACS Nano. **13**, 6224–6255 (2019)

Sun, H., Wu, L., Wei, W., Qu, X.: Recent advances in graphene quantum dots for sensing. Mater. Today. **16**, 433–442 (2013)

Tian, P., Tang, L., Teng, K.S., Lau, S.P.: Graphene quantum dots from chemistry to applications. Mater. Today Chem. **10**, 221–258 (2018)

Wang, Y., Hu, A.: Carbon quantum dots: synthesis, properties, and applications. J. Mater. Chem. C. **2**, 6921–6939 (2014)

Wang, J., Qiu, J.: A review of carbon dots in biological applications. J. Mater. Sci. **51**, 4728–4738 (2016)

Wang, J., Choi, H.S., Wang, Y.-X.J.: Exponential growth of publications on carbon nanodots by Chinese authors. J. Thorac. Dis. **7**, E201–E205 (2015)

Wang, R., Lu, K.-Q., Tang, Z.-R., Xu, Y.-J.: Recent progress in carbon quantum dots: synthesis, properties and applications in photocatalysis. J. Mater. Chem. A. **5**, 3717–3734 (2017)

Wang, X., Feng, Y., Dong, P., Huang, J.: A mini review on carbon quantum dots: preparation, properties, and electrocatalytic application. Front. Chem. **7**, 671 (2019a)

Wang, Y., Yang, P., Zheng, L., Shi, X., Zheng, H.: Carbon nanomaterials with sp^2 or/and sp hybridization in energy conversion and storage applications: a review. Energy Storage Mater. **26**, 349–370 (2019b). https://doi.org/10.1016/j.ensm.2019.11.006

Xu, Y., Liu, J., Gao, C., Wang, E.: Applications of carbon quantum dots in electrochemiluminescence: a mini review. Electrochem. Commun. **48**, 151–154 (2014)

Xu, D., Lin, Q., Chang, H.-T.: Recent advances and sensing applications of carbon dots. Small Methods. (in press) https://doi.org/10.1002/smtd.201900387

Yan, F., Sun, Z., Zhang, H., Sun, X., Jiang, Y., Bai, Z.: The fluorescence mechanism of carbon dots and methods for tuning their emission color: a review. Microchim. Acta. **186**, 583 (2019a)

Yan, Y., Gong, J., Chen, J., Zeng, Z., Huang, W., Pu, K., Liu, J., Chen, P.: Recent advances on graphene quantum dots: from chemistry and physics to applications. Adv. Mater. **31**, 1808283 (2019b)

Yang, S.-T., Ho, Y.-S.: Research performance and trends of fluorescent carbon nanoparticles: a science citation index expanded-based analysis. J. Nanopart. Res. **21**, 202 (2019)

Yu, H., Shi, R., Zhao, Y., Waterhouse, G.I.N., Wu, L.-Z., Tung, C.-H., Zhang, T.: Smart utilization of carbon dots in semiconductor photocatalysis. Adv. Mater. **28**, 9454–9477 (2016)

Yuan, F., Li, S., Fan, Z., Meng, X., Fan, L., Yang, S.: Shining carbon dots: synthesis and biomedical and optoelectronic applications. Nano Today. **11**, 565–586 (2016)

Yuan, T., Meng, T., He, P., Shi, Y., Li, Y., Li, X., Fan, L., Yang, S.: Carbon quantum dots: an emerging material for optoelectronic applications. J. Mater. Chem. C. **7**, 6820–6835 (2019)

Zhang, Q., Uchaker, E., Candelaria, S.L., Cao, G.: Nanomaterials for energy conversion and storage. Chem. Soc. Rev. **42**, 3127–3171 (2013)

Zhang, Z., Zheng, T., Li, X., Xu, J., Zeng, H.: Progress of carbon quantum dots in photocatalysis applications. Part. Part. Syst. Charact. **33**, 457–472 (2016)

Zhou, J., Zhou, H., Tang, J., Deng, S., Yan, F., Li, W., Qu, M.: Carbon dots doped with heteroatoms for fluorescent bioimaging: a review. Microchim. Acta. **184**, 343–368 (2017a)

Zhou, Y., Sharma, S.K., Peng, Z., Leblanc, R.M.: Polymers in carbon dots: a review. Polymers. **9**, 67 (2017b)

Zhu, S., Song, Y., Zhao, X., Shao, J., Zhang, J., Yang, B.: The photoluminescence mechanism in carbon dots (graphene quantum dots, carbon nanodots, and polymer dots): current state and future perspective. Nano Res. **8**, 355–381 (2015)

Zu, F., Yan, F., Bai, Z., Xu, J., Wang, Y., Huang, Y., Zhou, X.: The quenching of the fluorescence of carbon dots: a review on mechanisms and applications. Microchim. Acta. **184**, 1899–1914 (2017)

Zuo, P., Lu, X., Sun, Z., Guo, Y., He, H.: A review on syntheses, properties, characterization and bioanalytical applications of fluorescent carbon dots. Microchim. Acta. **183**, 519–542 (2016)

Index

© Springer Nature Switzerland AG 2020
Y.-P. Sun, *Carbon Dots*, https://doi.org/10.1007/978-3-030-41184-8

The manufacturer's authorised representative in the EU is Springer Nature Customer Service Centre GmbH, Europaplatz 3, 69115 Heidelberg, Germany. If you have any concerns regarding our products, please contact ProductSafety@springernature.com

Printed and bound by CPI Group (UK) Ltd, Croydon, CR0 4YY
22/09/2026
02221060-0004